Agricultural Engineering
Fundamentals and Applications

Agricultural Engineering
Fundamentals and Applications

— Editor —
Dr. S.N. Yadav
Associate Professor
College of Agricultural Engineering and Post Harvest Technology
Central Agricultural University
Sikkim

BIOTECH
2017
BIOTECH BOOKS®
Delhi - 110 035

2017
ISBN 81-7622-222-4
ISBN 978-81-7622-222-8

Published by : **BIOTECH BOOKS®**
1123/74, Deva Ram Park, Tri Nagar,
DELHI – 110 035
Phone: 27383999, (M) 09868812760
E-mail: biotechbooks@yahoo.co.in

Showroom : 4762-63/23, Ansari Road, Darya Ganj,
NEW DELHI - 110 002
Phone: 23245578

Printed at : Chawla Offset Printers
New Delhi – 110 052

PRINTED IN INDIA

Preface

Agricultural engineering, involving the application of engineering principles to agriculture, is one of the latest trends emerging in the arena, given an impetus to by the need for agricultural production to meet the growing demands of an ever increasing world population. Agricultural engineering entails the practice of such endeavours which include the designing of farm technology, food and bioprocessing technology, the search for biorenewable products. The field offers the prospect of not just better, scientific agricultural practice, but also sustainable agriculture.

The current text, as the title suggests, has been designed as an introductory text which undertakes a through exploration the concepts and practices which define agricultural engineering. Delineating the principles, technology and techniques which lend shape to this science, the text brings within purview the advantages and challenges offered by agricultural engineering, especially in the current global scenario where food prices have risen dramatically owing to insufficient agricultural production, Giving space to a thorough discussion of current trends in the field too, the text incorporates critical perspectives too, which aim to enhance the reader's understanding of the basics of the science of agricultural engineering.

S.N. Yadav

Contents

Preface v

1. Importance of Farm Mechanisation 1
2. Scope and Progress of Farm Mechanisation 13
3. Agricultural Engineering and Technology 26
4. Farm Power Sources 49
5. Impact of Mechanisation on Agricultural Productivity 66
6. Agricultural Machinery Industry 91
7. Custom Hiring of Agricultural Machinery 115
8. Long-Term Mechanisation Strategy 125
9. Safe Operation of Farm Machinery and Equipment 137
10. Tools for Tillage and Seebed Preparation 168
11. Harvesting Tools and Equipment 195
12. Temporary Fencing Techniques 215
13. Future Requirements of Agricultural Mechanisation 227

Bibliography 259

1

Importance of Farm Mechanisation

Mechanisation in Indian agriculture started with the establishment of the Central Tractor Organisation (CTO) mainly for land reclamation and development, mechanical cultivation and reduction of sanccharum spontanium (Kans). The production of irrigation pumps and diesel engines started during 1930s. The manufacture of tractors and power tillers started in 1960. Since then by the virtue of its inherent edge over the conventional means of farming, agricultural mechan-isation has been gaining popularity.

The increased use of farm machines has found expression in the phenomenal expansion of cropped area and cropping intensity and the country's agricultural production on all fronts. The shift has also helped in diversification of agriculture from conventional crops to commercial crops. The programmes of farm mechanisation have resulted in adoption of farm power and machinery such as tractors, power tillers, combine harvesters, irrigation equipment, plant protection equipment, threshers, improved implements and hand tools.

From 1986-87 to 1999-2000, there had been in general a rising trend in production and sale of different types of agricultural machinery especially tractors, power tillers, pumpsets, threshers, plant protection equipment, seed drills etc. After which, there has been a slightly decreasing trend of tractors and power tillers mainly due to unfavourable weather conditions, e.g. drought conditions in many parts of the country. However the country still lags behind in terms of the available number of tractors per thousand hectares. In terms of density of tractors per thousand hectares, India not only lags behind the developed countries but also some of the developing countries of the world, especially a few of the neighbouring ones.

The trend of sale of tractors over the years indicates a rising acceptance of agricultural machines and equipment with the Indian farmers. The use of different types of farm machinery including hand tools, animal-drawn implements, threshers, plant protection equipment, diesel/electric pump sets etc. has been accelerating over the past several years; as a result of which the total power available per unit area on farms has also increased.

Farm mechanisation has seen a rather slow progress over the years. The demand of tractors, power tillers, combine harvesters, irrigation pump sets, diesel engines, has shown an increasing trend. The progress of farm mechanisation in terms of demand of agricultural equipment is estimated at about 1 to 1.5% per annum. According to the Indian livestock census, the population of 44 items of Agricultural Implements and Machinery during 1977 was 109.66 million. This went up to 126.85 million in 1982 and further to 154.38 million in 1987 and then decreased to 141.93 million in 1992.

Again during 1997, it increased to 180.57 million and further to 223.84 million during 2003 (55 items). Introduction and adoption of agricultural machinery in the recent past has mainly been confined to the northern States of India. However, with the increase in the irrigation facilities and modernisation of the cropping practices, the demand for agricultural machinery has shown an increasing trend in the southern and western parts of the country. The eastern and the northeastern States have been less responsive to adoption of agricultural machinery.

Farm Power is an essential input in agriculture for timely field operations for operating different types of farm equipment and for stationary jobs like operating irrigation equipment, threshers/shellers/cleaners/graders and other post harvest equipment. During the last 50 years, the average farm power availability in India has increased from about 0.25 kW/ha in 1951 to about 1.35 kW/ha in 2001. Over the years, the shift has been towards the use of mechanical and electrical sources of power.

While in 1951, about 97.4% farm power was coming from animate sources, in 2001 the contribution of animate sources of power reduced to about 18% and that of mechanical and electrical sources of power increased from 2.6% in 1951 to about 82% in 2001. While the farm power availability of Punjab in 2001 has reached to over 3.5 kW/ha, it is still less than 0.90 kW/ha, in many States like Orissa, Rajasthan, Himachal Pradesh, Jammu & Kashmir, Chhattisgarh, Jharkhand, Gujarat, Assam, Madhya Pradesh and Maharashtra. The power productivity relationship shows that those States having higher farm power availability/ha have higher productivity. The State of Punjab which is highly mechanised, uses about 30% electrical power and about 48% diesel engine power.

It is visualised that the additional requirement of food grains in future will be met, to a great extent, from Indo-Gangetic plains where the demand of tractors, power tillers and other machinery will continue to increase in future also. For increasing productivity of dryland agriculture which constitute about 66% of the cultivated area in India, timeliness in farm operations is essential especially for seedbed preparation and sowing operations for establishing good crop stand in deficient/receding soil moisture condition. In these areas also the demand of tractors/power tillers, seed drills/planters and other farm machinery on custom service will increase in future.

Those States where farm power availability is less than 0.90 kW/ha will continue to use animate source of power for tractive purposes for quite some time before they shift to mechanical traction on a larger scale. States in NEH Region, Himachal Pradesh, Jammu & Kashmir and Chhattisgarh because of their hilly terrain will continue to use human and animate sources of power for a long time.

In dryland agriculture, to take the best advantage of available soil moisture, it is necessary to complete tillage operation as quickly as possible. For this, we require high HP and high capacity equipment to complete the tillage and sowing operations quickly, preferably in a single pass.

High capacity Machines/Equipment are required for the following reasons:

- Tillage and Sowing operations have to be completed quickly, often in a single pass,
- To make the best use of the available soil moisture to sub-soiling and use of basin lister may be required to improve moisture intake of the soil.
- Deep tillage after harvest in dry soil conditions is recommended to prevent run-off from the first rain,
- Seeding in deep furrows may be required if the soil moisture recedes beyond the normal seeding depth.
- Dryland agriculture, often associated with poor productivity and use of animal draft power and conventional tools, can be modernised, provided relatively large tractors and matching equipment are promoted in these areas. It is anticipated that in due course of time, the use of draft animals in rainfed areas will decrease and mechanical power will be a common feature of dryland agriculture.
- The use of high power tractors will facilitate land forming for water harvesting and conservation. Seeing the present trend and considering the future demand of additional power sources, it is visualised that by 2020, the average farm power need in India will be about 2 kW/ha of which the share of animate source will be only about 5% and that of mechanical and

electrical power will be about 70% and 25% respectively.

Role of Power in Farm Operations

Power is needed on the farm for operating different tools, implements and during various farm operations. While mobile power is used for doing different field jobs, the stationary power is used for lifting water and operating irrigation equipment; operating threshers, shellers/ decorticators, cleaners, graders and for other post harvest operations. The mobile farm power comes from human, draught animals, power tillers, tractors and self propelled machines, where as the stationary power is obtained from oil engines (diesel, petrol and kerosene) and electric motors.

Availability of adequate farm power is very crucial for timely farm operations for increasing production and productivity and handling the crop produce to reduce losses. With the increase in cropping intensity the turn around time is drastically reduced and it is not possible to harvest and thresh the standing crop, on one hand, and prepare seed bed and do timely sowing operations of subsequent crop, on the other hand, in the limited time available, unless adequate farm power is available.

Similarly for precision farming, increasing area under irrigation, conservation tillage, straw management and diversification in agriculture, more power is required for water lifting and precision placement/application of agricultural inputs-seed, fertiliser, irrigation water, plant protection chemicals etc and meeting the requirements of diversified agriculture.

There has been close nexus between farm power availability and increased productivity. Those States where availability of farm power is more have, in general, higher productivity as compared to others as shown in Table 1. The variations in the trend of productivity in few States are because of the variations in crops grown and the rainfall pattern in those States.

Table 1. Farm Power Availability and Average Productivity of Food grains in India in 2001

Sl. No.	Name of the State	Farm Power Availability (kW/ha)	Food grain productivity (Kg/ha)
1.	Punjab	3.50	4,032
2.	Haryana	2.25	3,088
3.	Uttar Pradesh	1.75	2,105
4.	Andhra Pradesh	1.60	1,995
5.	Uttaranchal	1.60	1,712
6.	West Bengal	1.25	2,217
7.	Tamil Nadu	0.90	2,262
8.	Karnataka	0.90	1,406
9.	Kerala	0.80	2,162
10.	Assam	0.80	1,443
11.	Bihar	0.80	1,622
12.	Gujarat	0.80	1,169
13.	Madhya Pradesh	0.80	907
14.	Himachal Pradesh	0.70	1,500
15.	Maharashtra	0.70	757
16.	Rajasthan	0.65	884
17.	Jharkhand	0.60	1,095
18.	Jammu & Kashmir	0.60	1,050
19.	Orissa	0.60	799
20.	Chhattisgarh	0.60	799
	All India	1.35	1,723

(Source:

The trend of average farm power availability and food grain production in India is shown in Figure 1.

Important Institutions

With a view to developing human resources in agricultural mechanisation and quality upgradation of the farm equipment, four farm machinery training and testing institutes (FMTTIs) were set up by the government of India.

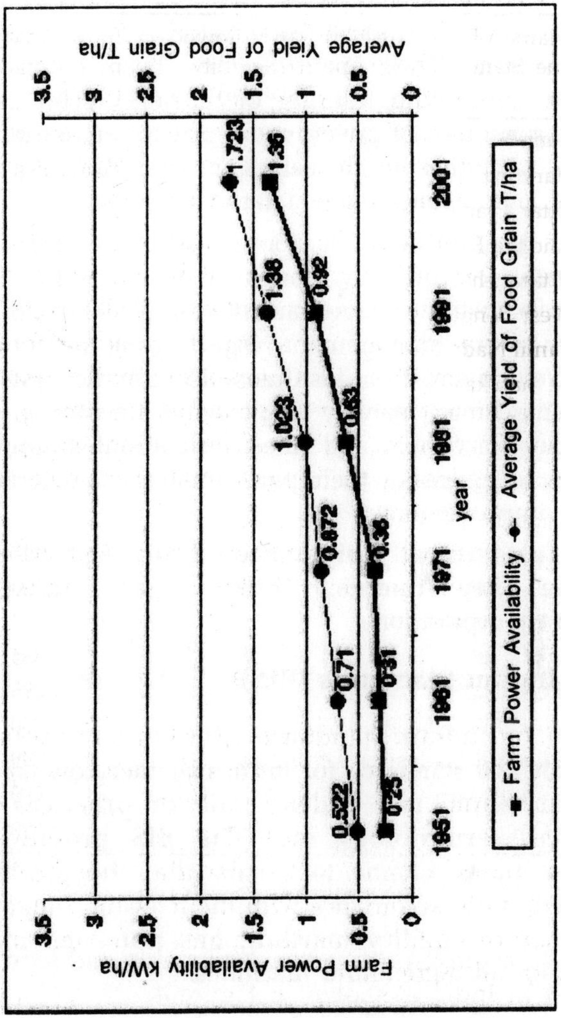

Figure. 1 Farm Power availability and average yield of Food Grains in India during 1951–2001

These are located as CFMTTI at Budni in Madhya Pradesh, NRFMTTI, Hissar in Haryana, SRFMTTI, Garladinne, Anantpur in Andhra Pradesh and NERFMTTI, Biswanath Chariali in Assam. These institutes have been assigned a mandate for conducting training of farmers, technicians, rural youth, nominees of government/private organisations and foreign nationals in the selection operation, repair/ maintenance and management of farm machinery.

Different types of short duration courses are organised round the year for different groups of beneficiaries. Since inception, these institutes have trained over 80,000 personnel in farm machinery management. Apart from the human resource development these institutes also conduct tests on agricultural machinery namely implements, tractors, power tillers, combine harvesters, and allied agricultural equipment with a view to assessing their performance characteristics and quality improvements.

ICAR Agricultural Universities State Agricultural Engineering Departments, States Agro Industrial Development Corporations.

Bureau of Indian Standards (BIS)

The Bureau of Indian Standards (BIS) has formulated approximately 300 standards for farm machinery on aspects like specifications, test codes, code of practices for maintenance, terminology etc. The BIS provides a certification mark scheme to certify that the products manufactured are in accordance with the relevant standards. The certification/quality marking has not been made compulsory on all agricultural machines.

Agricultural Machinery

Tractors

Tractors are manufactured in organised sector. Production of tractors began in India in 1961. Eicher Tractors was the first company to produce tractors from its unit at Faridabad in erstwhile Punjab. During 1961- 62, India produced 880 tractors. During 1986-87, the total population of tractors in the country was 80,369 and this reached to 2,78,556 during 1999-2000. From 1986-87 to 1999-2000, there has been in general a rising trend in production and sale of tractors after which there has been a slight decreasing trend in production of tractors, which has again picked up during 2003-2004.

Currently, there are about nineteen manufactures of tractors in India. Mahindra & Mahindra is the largest tractor producer in the country having a market share of over 25%. Tractors are mainly used in the States like Uttar Pradesh, Punjab, Rajasthan, Madhya Pradesh, Haryana and Gujarat. These six States together account for 78% of the total tractor population of the country. Punjab has the highest density of tractors in terms of gross cropped area.

Power Tillers

There are two main units in the organised sector and a few in the small scale sector manufacturing hand tractors (power tillers) in the country. Hand tractors were introduced in the country by importing a few models in 1961-62. Their indigenous production was started in 1965-66 by manufacturing 329 power tillers, it increased to 10,500 power tillers in 1995-96 and the same has touched 15,849 during 2003-2004.

Hand tractors are used for agricultural operations, transport purposes and stationary application for operating irrigation pumps, threshers, plant protection equipment etc. besides their suitability for wet land cultivation. The adoption of hand tractors is still at a very low level as compared to the four-wheel tractors in India.

Hand tractors presently being manufactured in the country, are in the range of 5 to 15 hp suitable mainly for wet-land cultivation. A light weight power tiller which could be useful for the hilly and terrace farming in the country would find place in the country.

Combine Harvester Thresher

Tractor-drawn as well as self-propelled combine harvesters for harvesting cereals and pulses have gained popularity among north Indian farmers. Eight units were earlier licensed to manufacture self-propelled combine harvesters, both in the organised and in the small scale sector. Production of combine harvester threshers has stagnated at around 300 to 500 per year, in recent times.

Agricultural Implements

The manufacture of agricultural implements in the country is reserved for the small scale industry. There are good number of registered small units and unregistered units manufacturing hand tools and implements. The States where agricultural implements are manufactured in India include Punjab, Haryana, Tamil Nadu, Karnataka, Gujarat, Uttar Pradesh, Madhya Pradesh, Andhra Pradesh, Bihar, Maharashtra etc. Though the country is self-sufficient in meeting the indigenous demand for the conventional agricultural implements, their quality is not up to the desired standard mainly because of limited know-how and financial resources of the small scale sector for quality control.

State Agro Industries Corporations (SAICs)

To provide broad-base technological services to the farmers by efficient supply of inputs including agricultural tools, implements and machinery and for extending custom hiring facilities, the government of India, through a centrally sponsored scheme promoted the establishment of 17 State Agro Industries Corporations (SAICs), one each in the major States of the country as joint ventures during 1965-70. The broad objectives in setting up of these corporations were:

- Manufacture and distribution of agricultural machinery improved implements and tools etc.
- To help persons engaged in agricultural and allied pursuits to own the means of modernising their

operations and making available necessary custom services for these purposes.

— Undertaking and assisting in the efficient distribution of inputs for agriculture.

— Promotion and setting up of Industries having a bearing on production, preservation and processing of agricultural products.

— Providing technical guidance to farmers and persons concerned with Agro-Industries, with a view to enabling them efficient management of their enterprises.

These corporations were not only allowed to augment their requirement by import of agricultural machinery but also two corporations in the States of Uttar Pradesh and Haryana set up tractor assembly plants and supplying them to all States and Union Territories in the country.

The initial role played by these corporations has been primarily responsible for development of the infrastructural facilities in the States at district and subdivision levels.

These Corporations also played a major role in supplying imported agricultural machines at reasonable prices and in building up the stock of spare parts and supplying the same to the users at competitive price throughout the respective States. Over a period of time, these corporations further expanded their areas and extent of activities.

Besides manufacturing and sale of agricultural tools and machinery including tractor, power tillers, oil engines, harvester, combines, threshers etc., these corporations also took up the projects for setting up rice milling plants, solvent extraction plants, fertiliser granulation plants, food and vegetable processing units etc. Initially State Agro Industries Corporations were set up to manufacture, demonstrate and promote agricultural implements and establish agro-processing plants in different States.

Over the period of time, however, the focus on these two programmes has been lost and the corporations are engaged in diversified activities not connected with the objectives with which these were established.

The programmes of the Corporations should be reoriented by the respective States so that these can effectively support the programme of agricultural mechanisation through supply of quality equipment, effective and intensive demonstration and training of mechanics and operators. These corporations should be headed by professionally competent persons.

References

Aggarwal, P.C. and Misra, M.S., "The combine harvester and its impact on labour: A Study in Ludhiana", *Indian J. Industrial Relations* 9(2), 1973.

Asian Productivity Organisation, *Farm Mechanization in Asia*, APO, Tokyo, 1983.

Chatha, I.S. and Grewal, S.S., *A study of tractorisation in Punjab*, Department of Economics and Sociology, Punjab Agricultural University, Ludhiana, 1991.

Datt, S. and Sundharam, K.P.M., *Indian Economy* S. Chand & Co. Ltd., 1998.

Gajja, B.L.; Kalla, J.C. and D.L. Vyas, *Determinants of tractorisation in arid areas of Western Rajasthan*, Agricultural Situation in India, 40(4):257-260, 1985.

Kumar, N. et al., *Farm mechanization and its impact on Punjab agriculture*, paper presented in I.S.A.E. Seminar report, Khalsa College, Amritsar, 1995.

2

Scope and Progress of Farm Mechanisation

Farm mechanisation has been helpful to bring about a significant improvement in agricultural productivity. Thus, there is strong need for mechanisation of agricultural operations. The factors that justify the strengthening of farm mechanisation in the country can be numerous. The timeliness of operations has assumed greater significant in obtaining optimal yields from different crops, which has been possible by way of mechanisation.

For instance, the sowing of wheat in Punjab is done up to the first fortnight of November. A delay beyond this period by every one week leads to about 3.5 quintals per ha decrease in the yield. This is also correct in the case of other crops and for other farm operations like interculture, irrigation, harvesting, threshing and marketing which need to be performed at appropriate time otherwise the yield and farm income is affected adversely.

Secondly, the quality and precision of the operations are equally significant for realising higher yields. The various operations such as land levelling, irrigation, sowing and planting, use of fertilisers, plant protection, harvesting and threshing need a high degree of precision to increase the efficiency of the inputs and reduce the losses. However, when such operations are performed through indigenous methods, their efficiency is reduced. Thirdly, the time taken to perform sequence of operations is a factor determining the cropping intensity. So as to ensure timeliness of various operations, it is quite inevitable to use such mechanical equipments which have higher output capacity and cut down the number of operations to be performed. This has helped in increasing area under cultivation and increase in cropping intensity.

Higher productivity of land and labour is another factor, which clearly justifies farm mechanisation. Not only the output per hour is more, the total labour requirement is also reduced. The displaced labour may of course be absorbed in the other alternatives created by the increased mechanisation such as manufacturing, repair and service shops and the sale services. Thus, it only results in the shifting of the labour from one vocation to the other. As production increases with mechanisation of the farm operations, it creates a good scope for commercialisation of agriculture.

Normally, there are good chances to reduce the cost of production if farm operations are mechanised as it saves labour, both human and bullock. In the absence of mechanisation, the ever-increasing wage rate of human labour and cost of upkeep of draught animals could have increased the cost of production much higher. Further, large scale production means less per unit cost on the farms. The use of farm mechanisation enlarges the employment opportunities both on farms and in nonfarm sectors through increase in area under plough, multiple cropping, development of agro industries and related services.

Farm Mechanisation Constraints

It is true that farm mechanisation has shown good results as of raising the agricultural production and improving the standard of living of cultivators within very short period. But a number of arguments have been advanced against farm mechanisation such as:

— Small size and scattered holdings of the farmers stand in the way of mechanisation. As a result of this, farm machinery generally remains underutilised,causing higher hourly cost of operation of machine.

— Majority of small cultivators are poor who are not in a position to purchase the costly machinery like tractors, combine harvesters etc.

— The use of tractor operated machinery may render some of the draft cattle population surplus. Studies under AICRP on Energy Requirement indicate that tractor owning farms do use draft animals for certain jobs. Likewise farms using animate sources of farm power, use tractor on custom service for certain jobs.

— The farm machinery require comparatively larger farm for economical use. Mechanisation may lead to structural change in agriculture in respect of the occupational distribution in the rural economy. No doubt, the increasing farm mechanisation is going to increase employment in secondary and tertiary sectors but it does displace labour in farm operations.

— Lack of proper knowledge of farmer to purchase farm machinery, operate and maintain it properly leads to wrong choice, makes it uneconomical and risky too.

— There is great shortage of diesel in the country as a whole. Thus, to use so extensive oil based farm machinery is not desirable.

— The lack of repair and replacement facilities especially in the remote rural areas is another hindrance in efficient small farm mechanisation.

— Due to the seasonal nature of the agriculture, the farm machinery remains idle for much of the time. Thus, idle machinery means unnecessary high costs unless proper alternate use of such machinery in the off-season is made.

Mechanisation Progress

The traditional farm tools and implement mainly relied on use of human and animate power. Improved farm tools, implements and machinery, which use both animate and mechanical power were devised from time to time. The average size of farm holding being small, animate power is widely used in many parts of the country. Mechanical power is making its impact in Indian agriculture with steady increase in land and labour productivity.

The traditional animal operated country plough although give low output and require higher number of field operations are still being used by majority of the farmers. Animal drawn cultivator and puddler have gained popularity showing an annual growth rate of 3.11% and 7.93% respectively due to higher output and better quality of work.

Improved implements such as M.B. plough, puddler, disc harrow, peg tooth harrow, spring tine harrow and patella harrow, being more efficient, have been adopted. Use of sowing/ planting devices for line sowing have also shown a growth rate of 6.5% as it helped the farmers in better management of costly inputs of seed and fertilisers.

The growth in the number of sprayers and dusters for plant protection has also been significant. The farmers are increasingly using power sprayers and dusters. The estimated population during 1995-96 was 0.25 million and was about 0.31 million in 2000. The number of draught animals has shown decline during the past few decades as a consequence of farm mechanisation and high cost of animal upkeep.

Irrigation is one of the major energy-intensive operation. With the increase in gross irrigated area, the number of irrigation pumps has swelled from 20 thousand in 1950-51 to about 12.51 million in 2000-01. Electric pumps are preferred than diesel engine operated pumps due to lower cost and higher energy-use efficiency.

Therefore, the growth rate of electric pumps was 8.20% as against 4.89% in diesel pumps. One of the major problems has been the lack of adequate and timely availability of electricity in rural areas due to which the diesel engines are kept as standby source of irrigation.

Tractor is the basic machine on which most of the farm mechanisation depends. The number of tractors was just 8000 in 1950-51, which steeply went up to 2.64 million in 2000-01 showing a growth rate of 10.3%. Power tiller was introduced in the country in the sixties, but could not gain popularity like tractor due to its limitation in the field and on the road. The power tillers are being used presently in rice and sugarcane producing areas of Tamil Nadu, Andhra Pradesh, Kerala, Karnataka, West Bengal, Bihar and Maharashtra.

Prime movers for irrigation are also used for operating threshing, chaff cutting, cane crushing equipment. Tractors are also used as stationary power source for such purpose. Use of power thresher, especially for wheat crop, became very popular in the seventies even among small farmers.

Employment and Farm Mechanisation

The use of different machines in agriculture has different types of impacts on employment of human and bullock labour. On the basis of data collected under the cost of cultivation scheme, the use of bullock labour has gone down significantly on the three major crops in Punjab as a result of fast mechanisation. Although the cost of human labour has gone up but there is considerable decline in the quantity of labour use.

Aggarwal and Mishra reported an estimated displacement of casual labour by cost of combine harvester in Ludhiana district to the extent of 22 mandays/ha. Rao studied the relative cost of traditional and mechanical methods of harvesting and threshing wheat. It was reported that harvester combine displaces labour on a large scale and was costliest from social point of view.

Mishra & Sundram assessed the cost and benefits of harvester combines. It was estimated that the use of harvester combine resulted in saving of about 37 mandays of unskilled labour per ha. The study also compared the cost of harvester combine with other alternative technologies and found it to be unprofitable unless farmers take third crop in a year.

Here, it is worth mentioning that it is wrong to say that all sorts of mechanisation are unjustifiable. For example, the use of pump sets has established the fact that labour displacement by them is more than compensated by the increased demand for labour in the field of irrigation and other operations. Thus, the question of farm mechanisation and unemployment is basically concerned with the use of tractors, threshers, combine harvesters etc. However, these machines bring timeliness and remove drudgery for farm operations, reduce unit cost of production in proving competi-tiveness.

Economics of Farm Mechanisation

It is of utmost importance to examine whether the use of machines has been economical or not. On the basis of a study covering 203 farmers having 218 tractors in different districts of Punjab it was brought out that the total use of the tractor, which on an average came out 397 hours per annum is much less than the possible extent of 1000 hours. The cost per hour turned out to be very high due to high fixed cost, which can be reduced by increasing the hours of working of the tractor.

The committee set up by the Planning Commission observed that harvester combines were generally demanded by big cultivators but it displaced a large number of agricultural labour in the harvesting season when the opportunities of employment in agriculture were higher for them. The committee justified the use of harvester combine only if their contribution to production arising out of saving in grain from vagaries of weather and shattering and from multiple cropping and change in cropping pattern was substantial.

Asian Productivity Organisation recognised time saved, freedom from over burdened work, improvement in social status, increase in overall production, timeliness of operations, reduction in cost, increase in the number of croppings and adoptions of inter-cropping as gains. Increased debt, cost of fuel and repair, unemployment, disparity in income were considered as losses due to farm mechanisation.

Inter-regional Variation

The annual growth of tractors in Punjab was 24.57% in the period 1961-66 which increased to 45.75% during 1966-72 and again declined to 16.49% during the period 1972-77. Tractor density, which was quite low being 1.31% prior to Green Revolution, moved up to 18.03% in 1976-77.

Umesh and Mathew studied the growth in production, imports and sales of tractors in Indian agriculture over the period 1960-61 and 1987-88. The study revealed that tractorisation had proceeded much faster in states such as Punjab and Haryana, which have gone through green revolution while many other states had not made much progress.

The district-wise distribution of tractors for the year 1987-88 was also examined by Chatha & Grewal. The study revealed that tractor population in different districts was quite skewed and it varied between 37.73 and 84.61 per thousand hectares of cultivated land in Gurdaspur and Ludhiana districts respectively. Singh revealed that number of tractors in Punjab increased from 10,636 in 1965-66 to 2,75,000 in 1990-91.

Sisodia found that there had been substantial increase in the population of tractors in the state of Madhya Pradesh and the number, which was 1,311 in 1956-57, increased to 3,544 in 1969-70. District-wise analysis showed that tractor population recorded significant increase in almost all the districts, and seven districts accounted for more than 50% of the total tractor population in the state during 1969-70.

The number of tractors per 10,000 acres of gross cultivated land increased steadily from 0.30 in 1956-57 to 0.70 in 1969-70.

The tractor density in the state increased from 10 per thousand hectares in 1971-72 to about 60 per thousand hectares in 1988-89. The inter-district tractor concentration revealed that the intensity of tractor population was largely correlated with the availability of irrigation facilities and the districts with more intensity of irrigation had higher concentration of tractors.

Rank correlation analysis indicated that agricultural productivity and irrigation were important determinants of tractorisation. The farm mechanisation is dependent mainly upon the size of operational holding, land topography, availability of credit facilities and per hectare profitability which in turn is affected by per ha yield, cropping intensity, market prices etc.

For the country as a whole, there were 63.47 tractors per thousand of cultivated hectares or 45.91 per thousand of cropped hectares in 1999-2000. The intensity of tractors in the hilly areas such as Sikkim, Tripura, Mizoram, Arunachal Pradesh, Assam was low because of operational problems of farm machinery and low crop yields.

Demand Projections

A single equation model was used by National Council of Applied Economic Research studied the demand for wheeled tractors in the country during the fifth plan period.to estimate the stock demand of tractors for 1973-74 and 1978-79. The variables tried in the model were relative price of the tractor, irrigated area, agricultural production and gross cultivated area. The study projected the stock demand level between 2.42 and 2.44 lakh tractors. Further, the annual demand for 1973-74 as 38 to 40 thousand units was estimated rising to about 79 thousand by 1978-79. Sharma projected that the demand for tractors in India to be 2258 thousand in 2024-25.

Singh and Jain fitted econometric model to estimate the demand for tractors. They found that the agricultural productivity and the labour force were found to be the important variables in the estimation of demand for tractors in the Punjab state. However, the influence of variables such as price of the tractors and irrigation was low. It was estimated that the stock of tractors in Punjab by the end of the century would vary between 1.70 to 2.40 lakhs.

The annual absorption of tractors could lie in the range of 10,000 to 16,000 which was considered quite compatible with the growth of labour force and agricultural productivity.

Chatha and Grewal analysed the growth pattern of tractorisation for the previous period and used it for estimating the future demand for tractors in Punjab. It showed that demand for tractors in the state during the year 2000-01 and 2030-31 would be 4.10 lakhs and 5.60 lakhs respectively.

The saturation point of the demand for tractors would be reached sometime during the period 2010-11 and afterwards only the replacement demand would continue which worked out to be around 25 to 32 thousand tractors annually.

Use of Tractor

The trend in farm mechanisation in India and the variables associated with it. It indicate that there was dramatic rise in the number of tractors during the period of mid-sixties due to Green Revolution.

Tractorisation was found to be positively correlated with variables such as percent of double cropped area to percentage of net area sown, percentage area irrigated to the net area sown, percentage area under high-yielding varieties to the net area sown, percentage area of holding with more than twenty hectares to net area sown, wages of agricultural labour and annual growth rate of agricultural output.

However, there was negative correlation of tractors with agricultural labour per 100 acres of net area sown and working animals per hundred acres of net area sown.

It was observed that tractorisation was significantly associated with higher level of high-yielding varieties and

HYV's were positively correlated with irrigation. Thus the interaction between tractors, high-yielding varieties and irrigation had led to the observed association between tractors and rate of growth of agricultural output. Sandhu examined the growth and determinants of tractorisation in Punjab. The study revealed that the number of tractors in the state increased 12 times over a period of about one and a half decade i.e. 1961 to 1975.

Rank correlation of tractor density with some important variables was established and found that the variables such as rural literacy, irrigation and cropping intensity, whose significance in determining tractorisation was not clear under static technology, because important in the context of dynamic technology.

Singh studied the determinants of tractorisation in Punjab using both rank correlation analysis as well as functional analysis. The study concluded that variables such as farm size distribution, agricultural productivity, irrigation, rural literacy, wages of agricultural labour were positively correlated with tractor density. However, there was a negative correlation of tractor density and tractor price, labour density and draft animal density.

Policy Measures

Mechanisation of Indian agriculture is the need of the time but its use has to be viewed from angles of unemployment problems of human and animal force and vast majority of small and uneconomical farms. To rationalise the existing farm machinery is important as seen from the high variability in inter-regional distribution, it is imperative to suggest some policy issues such as;

— The total use of the farm machinery, which on an average is much less than the prescribed norms. Therefore, there is need to enhance its productive use as a result of which, the fixed cost can be reduced significantly. The use of the machinery for social purposes, which is otherwise unproductive and thus needs to be minimised to the possible extent.

— The use of the tractor for custom hiring should be encouraged through legislative measures. The various problems faced by the farmers in this process such as uneven distribution of tractors, uniform cropping pattern in the area, social symbol attached to the ownership of tractors and the payment problems should be further probed into for their minimisation.

— The distribution of tractors from area to area should be normalised so that the custom hiring is increased. Further, the small farmers having operational holding of less than 4 hectares need to be discouraged to own the tractors unless they have substantial grounds to make economical use of it. However, to solve the problems of mechanisation of smaller holdings, the possibilities can be:

  — Cooperative management of farm machinery;

  — Financing of second hand tractors for small farmers;

  — Extension services to advise the suitability of various makes, models and horse powers for different size of operational holdings;

  — Devising smaller machinery suitable for small farms which constitute the vast majority of farmers but the machinery has to be effective and less costly.

— Suitable crop planning on the farm and in the area can be quite useful to gainfully employ the tractor. The monoculture of crops and varieties in the area results in high seasonal fluctuation in its use.

— There is need to revamp so that the land lease market is encouraged and the size of the tractor operated farms is increased to justify the existing farm machinery.

— Technical know-how should be provided to the farmers with respect to appropriateness of farm machinery for the situation and for its proper use.

— A standardisation and quality-marking centres of farm equipment should be established in potential areas of the country.

— The farm machinery suitable for different types of soil and operations for important crops should be developed and manufactured.

— The useful farm equipment should be identified and imported to cut down time and expenditure on research.

— Training relating to farm machinery and equipment should be imparted to the farmers and artisans.

— Service units should be encouraged and mobile service units should be started. Adequate and soft loans for such purposes should be provided.

Most of the negative arguments against farm mechan-isation can easily be solved without much difficulty. But there is no dispute regarding the need to strengthen mechanisation to boost the agricultural production and productivity to stabilise the economy.

Production and productivity, on the other hand, cannot be enhanced with primitive and traditional operations. Thus, keeping in view all facts in mind, selective mechanisation is the need of the hour.

References

Aggarwal, P.C. and Misra, M.S., "The combine harvester and its impact on labour: A Study in Ludhiana", *Indian J. Industrial Relations* 9(2), 1973.

Asian Productivity Organisation, *Farm Mechanization in Asia*, APO, Tokyo, 1983.

Chatha, I.S. and Grewal, S.S., *A study of tractorisation in Punjab*, Department of Economics and Sociology, Punjab Agricultural University, Ludhiana, 1991.

Datt, S. and Sundharam, K.P.M., *Indian Economy* S. Chand & Co. Ltd., 1998.

Gajja, B.L.; Kalla, J.C. and D.L. Vyas, *Determinants of tractorisation in arid areas of Western Rajasthan*, Agricultural Situation in India, 40(4):257-260, 1985.

Kumar, N. et al., *Farm mechanization and its impact on Punjab agriculture*, paper presented in I.S.A.E. Seminar report, Khalsa College, Amritsar, 1995.

Lekhi, R.K. and Singh, Joginder, *Agricultural Economics*,Kalyani Publishers, New Delhi, July 2002.

Mishra, P. and Sundram, *Some aspects of the economics of harvest combines in Punjab*, Economic & Political Weekly, 10(39), 1975.

PAU, Ludhiana, Directorate of Extension Education, *Package of Practices for Rabi Crops* in Punjab, 2002.

Pathak, A.S., *Farm mechanization and productivity*, Productivity, 23(4): 411-417, 1985.

Planning Commission, Govt. of India, *Expert committee on combine harvesters*, Report (Mimeographed), 1975.

Sethi, K., *Optimum land use pattern and normative demand for tractors in a growing economy-a case study of Ludhiana district*, unpublished M.Sc. Thesis, Punjab Agricultural University, Ludhiana, 1992.

Sharma, B.M.; Goyal, S.K. and G.P. Sharma, "Input use efficiency on tractor and bullock operated farms in Aligarh district of Uttar Pradesh", *Indian Journal of Agricultural Economics*, 47(3): 487, 1992.

Singh, B., *Demand for tractors in Punjab*, unpublished M.Sc. thesis, Punjab Agricultural University, Ludhiana, 1979.

3

Agricultural Engineering and Technology

TRACTOR-OPERATED MACHINERY

Tractor Mounted luged wheel puddler

A tractor-mounted, lug-wheel puddler of 1880-mm width has been developed for the shallow-tilled saturated soils. Its preliminary tests have indicated its average operation speed of 2.1 km/hr and depth of operation at 126 mm, and its effective field capacity is 0.32 ha/hr and field efficiency is 80%. It cost of operation has been Rs 192/hr, including that of tractor (Rs 185/hr). The unit has been tested and found satisfactory for shallow puddling. The weight of the puddler is 220 kg and its cost is Rs 8,500.

Seven-row Till-plant Machine

A seven-row, tractor-operated till-plant machine has been developed and was field evaluated in kharif and rabi 2002-03 for sowing soybean and chickpea. Its performance has been compared with the conventional practice of crop establishment, i.e. tillage by tractor-operated, five-row duck-foot cultivator, followed by seeding by tractor-operated seed-cum-fertiliser drill. This new machine has showed better timeliness of sowing operation, with reduced fuel consumption and reduced cost of operation, and yields under 2 systems were comparable.

Pneumatic Planter for Vegetables

Tractor-operated pneumatic vegetable planter has been developed with the field capacity 0.5ha/hr and field efficiency of 72.5% for peas (Arkil variety) in farmers' fields. About 90% of the seeds were within ±5 cm of the desired spacing of 20 cm. Seed rate of 50 kg/ha with this was 15-20% lower than the rate usually used by farmers. Its cost is Rs 40,000, and cost of planting by this is Rs 265/hr.

Two-row Vegetable Transplanter

A tractor-mounted, 2- row vegetable transplanter developed at CIAE, Bhopal, has been tested for transplanting tomato, cabbage, cauliflower and brinjal seedlings in the farmers' fields. Its field capacity has been found at 0.10 ha/hr with the field efficiency of 65-75 %. And cost of machine is estimated at Rs 26,500. With the manual method, 30 man-days per hectare are required for transplanting tomato at 60-cm row-to-row spacing and 45-cm plant-to-plant spacing. The operation cost of the transplanter for tomatoes comes to Rs 2,050 per hectare for 300 hr of annual use, and for manual method, it is 2,400/ha.

Zero-till Seed-cum-fertiliser Drill

Trials on tractor mounted, zero-till seed-cum-fertiliser drill were conducted for sowing wheat, covering 20.5 hectares. The machine gave an effective field capacity of 0.33 ha/hr at the forward speed of 3.2 km/hr. During trials, average soil moisture content (wb) was 23.9%. The average depth of sowing and field efficiency were 58.0 mm and 63.4% respectively.

MPKV Multicrop Planter

The tractor-mounted, multicrop planter has been designed and developed by incorporating salient features of animal-drawn Jyoti multicrop planter. This is suitable for groundnut, sunflower, chickpea, soybean, jowar and wheat. A plate in vertical plane with cells on its periphery is used for metering seeds for each row. A fluted roller is used for metering granular fertilisers. Depending upon the crop, number of rows can be adjusted from 5 to 9.

Furrow openers are spearhead type. Drive for metering mechanism is through a ground-wheel. The machine is mounted on a 3-point linkage system of the tractor of 35-hp or above. The initial cost of the machine is Rs 15,000. Its field capacity has been found 0.42 ha/hr with a field efficiency of 59%, and the germination percentage of the crop was 89%.

Strip Till Drill

Demonstrations on tractor-operated, strip till-drill for sowing wheat (PBW 343) were conducted covering a total of 7.7 hectares. At soil depth of 540-mm and soil-moisture content (wb) of 21.3 %, machine could give effective field capacity of 0.34 ha/hr, corresponding to the forward speed of 3.4 km/hr. In the delayed sowing, farmers could combine operations of tillage and sowing in a single pass by this machine.

Bed Planting of Wheat

This machine for bed planting consists of a drive-wheel, fluted roller-type seed mechanism and shoe-type furrow openers and furrowers for making bed. Its overall dimensions are 2,080 mm × 1,870 mm × 1,350 mm and can be operated with a 35-hp tractor. The machine has cup-feed-type fertiliser metering mechanism.

Flail-type Forage Harvester-cum-chopper

A flail-type, mower-cum-chopper for fodder-crop harvesting has been developed and commercialised. In a single operation this machine can harvest, chop and load chopped fodders like maize, bajra, oats in the trailer attached behind the machine. The machine consists of a rotary shaft mounted with blades to harvest crop, an auger for conveying cut crop, and cutters for chopping and conveying chopped fodder through outlet on to trailer. The blades on the rotary shaft are staggered in 3 rows of 13 blades on each row on the horizontal axis, perpendicular to the direction of the motion. After the crop is cut by blades, it comes to auger which conveys it to chopper unit.

The chopping mechanism having 3 blades, cuts crop into pieces, and chopped material is thrown out with high speed and is filled into hitched trailer. Its field capacity, labour requirement and fuel consumption varied from 0.44 to 0.52 ha/hr, 3.84 to 4.54 man-hr/ha and from 4 to 4.5 litres/hr respectively. The cost of machine is Rs 70,000 and cost of operation is Rs 1,113/ha.

Baler

Trials on tractor operated baler have been conducted. Its effective field capacity has been 0.82 ha/hr at the field efficiency of 87.2 %. Its bales (60 cm × 38 cm × 48 cm) output and average fuel consumption have been 1584 number/hr and 3.83 litres/ha respectively.

Post-hole Digger

Tractor-mounted post-hole digger has been field tested by the ANGRAU, Hyderabad with 30-and-23 cm diameter augers for making holes for tree plantation. The field capacity with 30-cm auger varied from 12 to 15 pits/hr for digging 230 holes to the depth of 60 cm with the field efficiency of 80%. Its cost of operation is Rs 19/pit in Red Chalka soil with 10% moisture (db).

Power-tiller-operated Machinery

Sweep Cultivator

It is suitable for interculture operations in crops having row spacing of more than 300 mm. This machine of 40-kg weight has overall dimensions of 930 mm × 1580 mm × 1,370 mm. It gives field capacity of 0.18-0.25 ha/hr for different crops. Trials on the power-tiller operated sweep cultivator were conducted in tobacco, covering total 1.3 hectares. For average tobacco-plant of 220-mm, cross-interculture operations gave weeding efficiency of 72-80 %. Its fuel consumption and labour requirement were 0.7 litre/hr and 4.5 man-hr/ha respectively.

Zero-till Drill Machine

This machine developed by the NDUAT, Faizabad, is suitable for 10-12-hp power-tiller. It can directly drill seeds and fertilizers without seed-bed preparation. It is suitable for wheat, barley, lentil, chickpea, pea, paddy etc. Its cost is Rs 10,000, and its cost of operation is Rs 420/ ha. It saves 68% in time, 85% in cost of operation compared to conventional practice, and increases yield by 6%.

Air-assisted Seed-drill

Power-tiller-mounted, air assisted seed-drill has been modified by providing stand for mounting blower assembly. One more furrow-opener with square-box holder has been fabricated and fitted on to the already fabricated tool-bar. For increasing seed carrying capacity, seed entry place in the air pipes has been modified. Final adjustments have been made to get blower outlet speed of 0.5-4 metres/second.

HPKV Multicrop Planter

It consists of a rotor-type metering mechanism with holes of different sizes on its periphery for different crops.

It can be used for two rows in maize and soybean and for three rows in wheat. It is operated at a speed of 2-2.5 km/hr in small terraces. Its effective field capacity and labour requirement have been found at 0.10 ha/hr and 19 man-hr/ ha respectively. The cost of the seeding with this planter has been 30-40% lesser than Kera method.

Orchard Sprayer

It consists of a horizontal triplex pump, working in an oil-bath. This pump gives discharge of 36 litres per minute and works at a pressure of 35 kg/ sq cm. It has main frame chassis, chemical tank, rubber wheels and operator's seat. This unit can also be conventionally used in pandal-type as well as Y-type of grape cultivation. For spraying in pomegranate and citrus, the booms one each on the right and one on left side, are made up from GI pipe. Its field capacity is 0.80-0.90 ha/hr and gives droplets of 200-250 µm and their density is 20-35 nos./sq cm. The height of the sprayer is 6 m and its operating cost is Rs 52/hr.

The orchard sprayer equipped with turbo-nozzles, developed at the MPKV, was used to conduct field tests in pomegranate and sweet orange. Its maximum number of droplets were in the range of 0-150µ with droplet density of 6-54 droplets/sq.cm. The values of number mean diameter and volume mean diameter varied from 51 to 73.31 and 51 to 217.94µ.

OUAT Groundnut Digger

A two-bottom, V-shaped blade-digger for groundnut has been designed and developed with 200-mm width and 38° rake angle of each bottom. The blade spacing can be adjusted according to the row-to-row spacing of the crop. Two bottoms have been provided with shank and tyne for increasing or decreasing shank height by loosening clamps. Inclination angles of the tynes to the ground level can be increased or decreased by rotating pipe-frame inside another clamp attached to hitch bracket. The digger (120° V blade) was tested at 2.1% soil moisture (db). Its effective field capacity has been 0.052 ha/hr with 78% field efficiency. Harvesting efficiency of the digger was 98%. Its cost of operation has been Rs. 1,375/ha compared to Rs 2,500/ha in the conventional practice.

Chipper Shredder

A power-tiller-operated chipper shredder for cotton-stalks and other agricultural waste materials like neem and subabul branches and leaves has been developed. Its shredder, flywheel cutter gets power through main clutch pulley of the power-tiller. The observed speed of the cutter flywheel has been 3,000 rpm. It was also evaluated for shredding coconut and chikoo branches. Its output capacity was 180-200 kg/ha, and shredded material size was 1-15 mm.

Two-row Cultivator for Biasi Operation

A self-propelled, 4.5-hp diesel-engine powered, 2-row cultivator for biasi operation for dry-seeded rice (broadcast/ row seeded) has been developed. When this machine was tested, 35 days after rice sowing under standing water, its effective field capacity was 0.06 ha/hr with a field efficiency of 80%. This field capacity is twice as much as that of animal-drawn, 2-bottom biasi plough and 3 times that of farmer's practice of single-bottom wedge plough. Estimated cost of this cultivator is Rs 50,000.

Riding-type (10-row) Rice-seeder

Self-propelled, riding-type rice-seeder (10-row) has been developed for sowing pre-germinated rice. It consists of a main frame of self-propelled (5-hp) unit of rice harvester mounted with rice-seeder (10-row). A wooden float has been attached to it and a seat has been provided to ride on the machine. The drive is taken from the ground wheel of the seeder to drop seeds.

Eight-row Rice Transplanter

It is a self-propelled, riding-type rice transplanter suitable for transplanting mat-type seedlings. It saves 65% labour and operating time 35-40% cost of operation, besides increasing yield 5-10% of the conventional method of manual transplanting. The costs of machine is Rs 120,000 and its cost of operation is Rs 1,000/ha. Mat type of seedlings helped in growing more number of tillers per hill. Its feasibility testing was done with Pant 12 rice in 2.35 hectares. Its field capacity was 0.13 ha/hr and field efficiency was 68%. Missing hills were 2.7% and floating hills were 0.6%. Number of seedling per hill were 3-5 at a 140-mm hill-to-hill spacing.

Power Weeders

CIAE Design

A self-propelled interculture equipment has been developed utilising chassis of 1-m self-propelled, vertical-conveyor reaper by replacing the present diesel-engine with a light weight petrol-start kerosene-run 1.1 kW engine having rated engine speed of 1,500 rpm (at the cam shaft). This is a light machine and can operate with 3 sweeps of 150 mm for weeding operation in crops sown at 300-350 mm row to-row spacing. If row spacing is 400 mm or more, only 2 sweeps of 150-200-mm size can be operated. A set of narrow wheels of 150-mm width has also been developed to facilitate operations of machine during weeding.

ANGRAU Model

This self-propelled rotary weeder is run with a 6.5-hp diesel-engine. Drive to rotary weeder is taken through two sets of V-belts and pulleys. A multiple-plate clutch is provided to cut-off drive to ground-wheel and weeder unit. The rotary weeding unit has two gangs of 300 mm each. Both gangs are mounted outside wheel tread of prime mover. In its feasibility trials, machine was operated at 60-cm row spacing and 120-cm spacing between pairs. At 30-40% soil moisture (db) and 30-40-mm depth of tilling, its weeding efficiency was 76% at 2.5 km/hr machine speed.

TNAU Model

It consists of a power unit equipped with 5.4-hp diesel-engine of 34-kg weight, power transmission gear-box, ground wheels, weeding unit handle and clutch. The width of coverage is 350 mm. The overall dimensions of 90 kg weeder are 2,100 mm × 640 mm × 1,170 mm. The machine can be used in maize, sugarcane, cotton, tapioca and grapes. The cost of equipment is Rs 55,000 and its cost of operation is Rs 770/ ha. After one pass with rotary weeder, a ridger can be attached for earthing-up. In cotton, field capacity, field efficiency, weeding efficiency and cost of operation have been found as 0.23 ha/hr, 84 % and 53 % and Rs 45/hr respectively and for sugarcane, these have been 0.116ha/hr, 59%, 38% and Rs 47/hr respectively.

PAU Walking-type Sprayer

The self-propelled, light-weight boom sprayer fitted with 5-hp diesel-engine consists of a spraying unit at the back. The spraying unit has a boom with 12 nozzles, spray pump, tank for liquid and a pressure gauge. Ground clearance of the machine is 500 mm. The boom height can be adjusted from 600 mm to 1,300 mm. The unit controlled by the operator is provided with two narrow wheels and one supporting wheel at the back. The unit has provision to adjust track width from 900 mm to 10,50 mm. The swath-width of the sprayer is 6,300 mm.

Its feasibility trials were conducted on 26 hectares for wheat. The unit operated at operating pressure up to 400 psi. The machine sprayed weedicide which gave field capacity of 1 ha/hr at the forward speed of 3 km/hr. The fuel consumption varied from 0.4 to 0.6 litre/hr. It could save 70-80% of labour and 40-50 % in cost of operation compared to knapsack sprayer.

Vertical Conveyor-reaper

This walking-type harvester for wheat, rice and safflower is suitable for cutting and windrowing crops. The engine power is transmitted to cutter-bar and conveyor belts through belt-pulleys. The prime mover of the machine is 5-6.5-hp diesel-engine. And the cost of the machine with prime mover is Rs 50,000.

Demonstrations were conducted with this for safflower crop covering 6 hectares. Its effective field capacity was 0.2 ha/hr and labour requirement was 40 man-hr/ha. Front line demonstration for wheat covering 3 hectares conducted at 2.7 km/hr forward speed, showed 0.22 ha/hr, 68% and 1.16 litres/hr effective field capacity, field efficiency and fuel consumption respectively. The working width, height of the cut and the average moisture content of the crop-stem were 1,140 mm, 54 mm and 12.7% (wb). The total losses during harvesting were 2.14%.

Animal-drawn Machinery

CIAE planter

It is suitable for planting groundnut, maize, pigeonpea, sorghum and other oilseed and pulse crops. It saves 64% on the operation cost compared to conventional method of sowing behind the country-plough with manual seed dropping. It costs Rs 6,500 and its cost of operation is Rs 390/ ha. Its field capacity vary from 0.12 to 0.15 ha/hr.

Stationery Machinery

CIAE Solar-tracking Device

Sun tracker developed for 450Wp SPV panel has been modified to be adopted for 900 Wp SPV panel to make it more reliable and user friendly. It consists of an electronic vibrator circuit to produce 38 electric pulses per minute, which activates an electromagnetic relay and ratchet and pawl to produce precise rotation of the shaft, mounted on the ball bearings. The tracker has been extensively evaluated by tracking panel (from morning to evening) for 700 hr of operation and has been found to give satisfactory performance.

The power output from the panel increased by about 30% under tracked condition (6000 Wh) as compared to non-tracked condition (4500 Wh) on the typical sunny day. The peak power output was observed to be 660 W (74% of peak installed capacity) during noon time on a typical sunny day in summer.

High-capacity Pigeonpea Thresher

A high-capacity pigeonpea thresher (10-hp) has been designed and developed for the crop-length of more than 750 mm at CIAE, Bhopal.

It consists of an automatic chainconveyor type feeding mechanism, a tapered spike tooth type threshing cylinder, a woven wire-mesh concave, two aspirator blowers, a shaker assembly and transport wheels. Stalks are broken into big pieces which can be easily used by farmers for domestic use. Threshing and cleaning efficiencies of this varied from 96.59 to 96.74% and 92 to 94.22% and The average broken grain, blown grain and sieve overflow were 1.21%, 0.31% and 0.13% respectively.

High-capacity Multicrop Thresher

The threshing capacity of the machine vary between 2.13 and 2.41 tonnes/hr. Threshing and cleaning efficiencies were 99-99.5% with total losses of 1.98-2.35%. Fuel consumption and cost of operation were 3-3.25 litres/hr and Rs 165-310 per tonne. Demonstrations for wheat (HD 2285) were organised. During trials threshing capacity, grain damage and straw size were 0.8-1.05 tonne/hr, 0.5-1% and 25-30 mm, respectively. Grain moisture and straw moisture (db) were 8-10% and 12-14%. The fuel consumption was 4-4.5 litres/hr.

Maize Dehusker-cum-sheller

Modified peg-tooth thresher has been developed for maize-crop. For better shelling efficiency varying height pegs have been provided in staggered fashion. The pegs have been placed in 6 rows with 6 pegs in each row. Sheller sieves of 12.5 mm are used. The front and rear concave clearances are 50 and 25 mm, respectively.

The concave grate has openings of 50 m × 50 m. The sieves of 12.5-mm size holes have been provided. Test trials conducted for maize sheller have showed grain-straw ratio and moisture content of grain (wb) as 4.2 and 12.1%. At threshing cylinder speed of 678 rpm and feed rate of 1,440 kg/hr, threshing efficiency was 99.26% and cleaning efficiency was 99.96%. The output capacity and broken grains varied 0.63-0.67 tonne/ha and 2.6-3.02%, respectively.

ANGRAU Sugarcane Leaf Stripper

The stripper equipped with 3-hp diesel-engine has been developed for estimated output capacity of 2 tonnes/hr. The equipment consists of 2 upper adjustable rollers and 2 lower fixed rollers. Rollers are covered with EPDM material on the circumference. Stripper is also provided with two brush carriers over which 4 nylon brushes are fixed at equal angles (90° each). The gap between two rollers is adjusted according to canes size; diameter of input rollers is kept lower than that of output rollers to enable canes to come out quickly after stripping leaves.

The speed of the brush holder is 2,000 rpm and of feed roller input and feed roller shaft is 1,000 rpm and 375 rpm. Sugarcane is stripped by passing cane between 2-hand rubber rollers. Two brushes with hand-nylon strings are provided to strip cane-leaves. During trials at Hyderabad, effective field capacity and fuel consumption were 1,300 kg/hr and 0.40-0.50 litre/hr, respectively and the breakage of canes, efficiency of machine and labour engaged were 12.15%, 65% and 5.

Ergonomic Studies

Anti-vibration devices for comfort of power-tiller and tractor operators: Vibrations if transmitted to body parts not only cause discomfort to operator but also reduce efficiency of the operator. Vibration isolators for engine, handle-bar and handle of the power-tiller have been developed and tested for their efficacy for attenuation of vibrations transmitted to operator. The isolators resulted in reduction of handle vibration by 50%. Besides these, vibration isolators have been developed for seat and tractor-trailer hitch points and have been tested for their efficacy. The isolator below the seat helped in reducing vertical vibrations transmitted to operator by 30-80% in different operations at speeds ranging from 2.5 to 4.5 km/hr.

Optimum locations for tractor-control pedals for Indian operators

Right and left leg strength data for 20 subjects were collected at 5 horizontal (35, 40, 45, 50 and 55% stature) and 6 vertical (10, 13, 16, 19, 22 and 25% strature) locations of the foot-pedal from the seat reference point (SRP), keeping lateral distance from midline 200 mm for each pedal combination. It has been observed that a 5th percentile operator could reach vertically up to 370 mm below and 810 mm in front of the SRP. The maximum leg strength of 655 N for right leg and of 613 N for the left leg was observed when pedal was located at 55% stature in front and 10% stature below SRP.

The optimum locations for brake and clutch pedals have been at the horizontal distance of 623 mm in front of the SRP, vertical distance of 296 mm below the SRP and 200-mm lateral from mid-line. It will be suitable for the 90% of the user population. The range of dimensions presented in IS: 12343 is very large and actuating force limits as given in IS:10703 are very high compared with leg-strength data of the Indian operators. Both these standards need to be revised in the light of the data generated to make the tractor work-place safe and comfortable for operators.

Post-harvest Management

Dust Emission in Dal Mills

A prototype of the cleaner equipped with suitable ducts, a forward curved centrifugal blower and a cyclone separator to arrest dust emission in dal mills has been developed. The diameter of the blower impeller is 300 mm with an inlet diameter of 150 mm. Its volumetric airflow rate is 22 m3 / min and air velocity at winnowing section is 10 m/s.

The blower housing has a diffuser angle of 5 degrees. Test results indicate that power requirement at load is 500 W and its input capacity is 900-1,000 kg/hr. The estimated cost of the cleaner with two prime movers (½ and 1 hp-electric motor) comes to about Rs 20,000. The screen effectiveness of the unit is 68-87%, and the purity of the cleaned grains is over 98 %.

Fermented Banana Beverage

A process for fermented banana beverage has been developed from banana pulp, water and sorghum flour in 1:4:0.83 ratio. A pre-grown, fresh-yeast culture suspension (Saccharomyces cerevisiae) was inoculated for fermentation. The average yield of the beverage has been 58% of the total mixture, which was equivalent to 65 bottles of 800 ml each.

The final product contains 5-8% alcohol with pH of 3.5, which is suitable for beverages. The beverage has sour-sweet taste and beer-type flavour. The bottled beverage can be stored safely for 45 days at the room temperature (30-35°C).

Jaggery and Khandsari Management

IISR horizontal cane-crusher

The IISR designed improved horizontal power cane-crusher gave maximum of 64% juice extraction at 3.7-mm roller-gap at 3.3m/min. roller speed. At this roller setting, power requirement for crushing 1 tonne of sugarcane was 7.9 kwh.

Double stage filtration system

To obtain quality product, clean juice free from adulterants, dust and any other foreign materials, a filtration system for juice has been developed. In this, first-stage filter is made up of stainless steel 304 of 20 gauge sheet. It contains 3-mm round-hole sieve and has 29 holes/sq inch. Similarly second-stage filter is also made up of the same material with 0.5-mm sieve. For cleaning from very small particles, of insoluble impurities, third filter of 500 micron size has been provided on the top of the juice-tank.

Digital thermometer developed for judging striking point in jaggery-making has minimized risk of charring and of sub-optimal condition to remove concentrated mass from the pan. The instruments costs Rs 800 to Rs 5,000.

Vegetable clarificant for sugarcane juice

Optimum clarification of juice with deola seed powder was achieved at 0.1 to 0.5% concentrations. Apart from being good in physical appearance, jaggery obtained was hard crystalline and light-yellowish because of the instrinsic properties of the seed powder for enhancing efficiency of juice clarification. The clarificant could be effectively used for preparation of syrup and vinegar.

3-pan furnace for jaggery manufacture

This consists of a combustion chamber with an improved grate design, a middle chamber for tapping heat from the flue-gases coming out of the combustion chamber and also to work as an air inlet chamber, and a flue-gases passage/ channel rectangular in shape, 3 juice-boiling pans, a chimney made of masonary for taking out fluegases, a bagasse feeding platform and an ash pit below the platform. The specially designed and fabricated juice-boiling pans were placed on the respective chambers.

The first 2 pans are circular with convex bottom having 11/4" and 1" thicknesses and the third one is known as the gutter pan, it is rectangular with convex bottom (1/2" thick) along the length. The middle juice boiling pan has 11/2" × 11/2" hole/opening in its centre which connects with 2" and 35" long g.i. pipe from the bottom of the pan for draining-out juice.

A gate valve has been provided at the end of the juice pipe. Also a protective cover of the thick m.s. sheet has been provided for this juice pipe. In additional to fuel-feed hole, the fresh air is supplied to combustion chamber enrouting middle chamber through two m.s. pipes (3") connected with two side openings of the chimney near its bottom and at the middle chamber. The chimney height is kept at approximately 12' with the base chamber of 5'3"×5'3"×2' and chimney of 2'9"×2'9"×12' with a hollow outlet opening of 1'3"×1'3". The test results have indicated that per batch approximately 20-25 kg of bagasse could be saved.

Jaggery in vacuum and nitrogen environment

Brix of jaggery obtained from sugarcane CoLk 8102 changed from 13.67 to 11.87, 12.3 and 12.0 when stored in the desiccator, nitrogen and vacuum respectively in 2 months. Similarly pol reading changed from 77.4 to 48.8, 72.3 and 70 g while reducing sugar increased from 5.7 to 8.5, 6.2 and 6.7% in desiccator, nitrogen and vacuum storage. Moisture content of jaggery changed from 7 to 16.5, 8.9 and 3.9% when it was stored in desiccator, nitrogen and vacuum.

Jaggery solution pH changed from 6.5 to 5.5, 5.9 and 6.2 and colour reading changed from initial value of 190 to 275, 220 and 252 in desiccator, nitrogen and vacuum stored jaggery. This suggests that packaging of jaggery in nitrogen environment will be more suitable.

Use of Plastic in Agriculture

At the CSKHPKV, Palampur, and the CIPHET, Abohar, plastic mulch (black polyethylene sheet of 100 gauge) has recorded 30-40% higher yield of strawberry over straw and eupatorium mulch. Yield-attributing characters like berry weight, length and diameter were also significantly higher with plastic mulch. Runners planted at 15-cm rowspacing produced significantly higher yield (25 tonnes/ ha)

Turning-red tomatoes stored in the perforated polybags with ethylene absorbent at room temperature had shelf-life of 10 days, and it increased to 18 days when tomatoes were stored at 18°C at the PAU, Ludhiana.

Mature green tomatoes with ethylene scavenger could be stored for 16 days at room temperature and 19 days at 18°C. The lycopene content and TSS have been found significantly higher for tomatoes packed in polybags without perforations. Tomatoes harvested at mature green or turning-red stage have better shelf-life.

Cotton Technology

The CIRCOT has developed a light weight gin. The weight is reduced by about 60%, keeping output at 5 kg seed-cotton per hour. This gin can be operated with a remote control as well. Heap-making machine for cotton: The CIRCOT has designed and developed a mechanical heap-maker that is able to throw cotton into well-formed heaps at the rate of 3,000 kg/hr. Efforts are on to increase heaping rate and bulk density of heap.

Highest Standard Count (HSC) is a single integrated index that provides an easy way to express quality of cotton and is considered a unique mean of expressing the maximum spinning potential of the given cotton.

The CIRCOT has developed an Artificial Neural Network model (ANN) using fibre properties such as 2.5% span length, uniformity ratio, micronaire value, bundle strength and percentage of mature fibres as input to predict highest standard count (HSC) value. This model is able to predict HSC within an error of ± 4.23 counts. Since the spinning industry is very familiar with the concept of count, understanding and using HSC as an index for characterising spinnability would not pose any problem and would be more favoured than the current Fibre Quality Index, which is used to indirectly assess spinnability of cotton.

Production of Xylanase

Xylanase is an enzyme that can biodegrade lignocellulosic substrates to useful end-products. Fungus Penicillium funiculosum could secrete xylanase simultaneously when cultivated for cellulose production using cellulose as the carbon-source.

The maximum yield of xylanase could be obtained 72 hr after incubation when substrate on which organism was cultivated had 1.0% cellulose and 0.25% peptone. The enzyme thus produced had maximum activity equivalent to 33.3 units/ ml at a pH of 6.0 and temperature 50°C that is considered optimum for use in textile finishing applications.

Angora Rabbit Hair Blends with Cotton

Angora rabbit hair with cottons like Suvin and DCH 32 have been blended. Since Angora fibres are shorter compared to wool, they cannot be effectively utilised through longstaple woollen and worsted spinning systems for manufacturing consumer acceptable textiles. This novel blending in cotton system is beneficial way of utilising rabbit hair.

Knitted fabrics produced from blended yarns are found to possess adequate bursting strength, whiteness, softness and bulkiness. Several innovative end-products like low-shrink knitted fabrics, single jersey and light weight soft feel women's wear have been prepared.

Effect of Lac Technology

Vegetables growing as an intercrop in the plantation of Flemingia semialata has showed significant improvement on the growth attributes of the lac-host plants such as height, basal girth, canopy spread and shoot length, and this will help to meet diversified needs of lac growers. Identified and collected rare variants of palas (Butea monosperma): Three rare variants of palas (Butea monosperma), with unifoliate flowers, yellow flowers and white flowers have been identified, collected and planted in the institute plantation for conservation of plant biodiversity.

Aleurityl hydraside was synthesized from methyl aleuritate by treating it with hydrazine hydrate in methanol. The compound showed antifungal activity. A sex pheromone of Helicovoerpa armigera (cotton bollworm), (Z)-9-Hexadecenal, was synthesized in quantity from aleuritic acid, the major component acid of the lac, with an overall yield of 20%.

New cyclic thioureide has been synthesized from aleuritic acid adopting simple reaction sequences. Shellac-based can-lacquer formulations for packaging. Various can-lacquer formulations developed have been applied on the aluminium foil by spray, with/ without suitable pigments, for possible use in packaging of sweets and confectioneries. The results obtained are quite satisfactory with regard to flexibility, attractive appearance and adhesion

Effect of Jute Technology

Because of acute shortage of retting water, farmers are compelled to ret jute-plants for fibre extraction in roadside ditches, ponds and canals. As a result, quality of the extracted fibres is reduced and farmers fail to get the remunerative price from their produce.

With the NIRJAFT technology, retting is effected in just water-soaked conditions by the application of a specific fungus belonging to Sclerotium group, thus reducing substantially water requirement. Bench-scale trials have found that with this retting could complete in a shorter time than traditional retting.

The NIRJAFT has worked on blending of coir and jute successfully in jutespinning system in 20:80 ratio. Better results could be obtained when coir fibre prior to blending was steamed under 15 psi pressure. The blended products are much better in productivity, strength and extension properties than 100% coir products.

Tasar waste has been blended with varying proportions of ramie ranging from 100 to 25%. Spinnability and regularity of the blended yarns (51-104 tex) improved due to higher fineness of tasar, which provides higher number of fibres in the same cross-section of yarn. The blended yarns have been found suitable for manufacturing garments, apparels and decorative fabrics.

About 20% sisal could be blended with 80% jute. Strength and durability of these blended yarns increased significantly. And these yarns can be used for making scrubber, matting and floor covering.

Energy in Agriculture Technology

Crop residue conversion to liquid fuel. Pretreatment of paddy-straw (400 micron) with 2% NaOH for 1 hour at 15 psi pressure in autoclave could result in 88% delignification and 100% increase in cellulose content.

Hydrolysis of alkali-treated paddy-straw with 75% (w/v) sulphuric acid resulted in release of 30% (w/w) sugars. And hydrolysis with crude culture filtrate of Trichoderma reesei (5FPU/g) and commercial cellulose (1.5% v/v) resulted in 48 and 81% saccharification of cellulose. Fermentation of hydrolysates of alkali-treated paddy-straw obtained after treatment with acid, crude culture filtrate and commercial enzyme and supplementation with 1.24% yeast nitrogen-base using Saccharamyces cerevisiae (1% w/v) could produce 78, 139 and 224 ml ethanol/kg paddy-straw.

An open-core gasifier of 700-mm internal diameter and 1,860-mm height has been developed. This has been fitted with a continuous ash removal mechanism. Outer surface of the reactor has been provided with glasswool insulation to minimize heat loss. The gasifier has been tested with bagasse briquettes of 30-mm diameter and with an air-flow rate of 200 m^3/hr, which could generate producer gas at the rate of 257 m^3/ hr.

The flame temperature of 657°C could be obtained. The system has been installed at the site of the sugarcane-farmer for jaggery preparation. Burner efficiency has been 52% as compared to 16% for the conventional system. The system could save 40% of fuel and 45% in time compared to conventional system for jaggery-making.

An unglased solar air-heater has been developed, which also works as the roof of the building. This heater has been made using hollow black colour PVC section in one mm wall thickness. One module of the heater has 1.75 m^2 surface area (2,730 mm × 300 mm × 25 mm). Single module of the heater was evaluated and a temperature rise of about 15°C was recorded. Thereafter, a large system measuring 21m^2 surface area, which formed roof of the small shade, was erected and its performance was evaluated during summer and winter.

The increase in air temperature varied between 18.9° and 13.2°C during summer (solar insolation 917 to 1069 W/m^2) and 16.6 to 10.2°C during winter (solar insolation 750-930 W/m^2), and air-flow rate varied between 441 and 1,075 m^3/hr.

The heater did not shown any discolourisation or wear and tear during 18 months of its exposure to outdoors. It costs around Rs 1,000 per m^2.

A SPV system for operation of a top-opening, 73-litre capacity DC refrigator has been developed. This refrigerator system was used to store animal vaccines and other medicines in the village veterinary clinic. Performance of the system has been found satisfactory over 9 months.

Keeping in view the requirement of the rural industry, a 80-litre capacity, top opening refrigerator mainly to store single products has been designed and developed. This refrigerator is equipped with Danfoss make DC compressor of 12 V/24V. And thermostat with seven settings enables user to obtain a wide range of temperature varying between -4.6 and 1.5°C. Its daily energy consumption varies between 450 and 875 Wh.

The cylindrical gasifer (700-mm diameter and 1,260-mm height) has been provided with 4 air-tiers located above the grate at the intervals of 90° at the periphery. Gas outlet of diameter of 300 mm (located 800 mm above grate) is fitted at an angle of 30° on the periphery of the gasifier with chimney at the height of 126 mm from the grate. An ash removal system has been provided for agitating fuel-bed and removal of ash from the grate. The gasifier also has a light-weight, fire-brick inner insulation of 50 mm.

The gasifier has performed satisfactorily with a thermal output of 106.7-121.3 kW. Its biomass consumption rates ranged between 35 and 40 kg/hr for wood-chip and 24-37 kg/hr for soybean-straw. The respective grate temperature was 1258-1380°C for wood-chips and 1154-1234°C for soybean straw. The provision of insulation could increase the grate temperature by 1.6 to 3.9%. The outer surface temperature of the gasifier near the grate remains 65-98°C, with surface temperature decreasing with the height to around 50°C at 300 mm from the gasifier-top.

Use of Animal Energy

The single-row, zero-till drill developed requires a draught of 35 kgf and its field capacity has been found at 0.03 ha/hr. The two-row, zero-till drill has a field capacity of 0.05-0.06 ha/hr and its draught requirement has been 45+10kg for light soils and 75+10kgf for heavy clay soils. Zero-till drill had resulted in 5% increase in yield as compared to conventional method.

Pantnagaradjustable collar harness and Allahabad harness have been modified to increase yoke-contact area for providing comfort to animals during work. The increased output has been utilised for operation of 100-150-mm plough with a single animal harness using buffaloes with a work-rest schedule of 4 hours work and 7 hours rest for the field preparation during rabi.

The animals could operate plough without showing any additional fatigue symptoms. The collar harness intervention has resulted in 20-25% more power output. The contact area could increase to 17% in modified harness. The improved implements matching draught capacity of animals under farmer's field could increase command area from a pair of buffaloes by 22-40% in various crop rotations, when buffaloes were used as a power source in tarai region.

The device consists of an improved design of bull-gear coupled to generator for electricity generation. The bull-gear consists of spur and helical gears and all gears are arranged vertically. The gear and shaft and key are made of EN 19 and EN 8 steel for its mechanical strength, longer life and also for amenability to surface hardening. The magnification ratio of a bull-gear has been 310-08. The bull gear coupled to 1 KVA alternator operating at 750 rev./min could give the power output ranging from 500 to 750 watts.

References

Aggarwal, P.C. and Misra, M.S., "The combine harvester and its impact on labour: A Study in Ludhiana", *Indian J. Industrial Relations* 9(2), 1973.

Asian Productivity Organisation, *Farm Mechanization in Asia*, APO, Tokyo, 1983.

Chatha, I.S. and Grewal, S.S., *A study of tractorisation in Punjab*, Department of Economics and Sociology, Punjab Agricultural University, Ludhiana, 1991.

Datt, S. and Sundharam, K.P.M., *Indian Economy* S. Chand & Co. Ltd., 1998.

Gajja, B.L.; Kalla, J.C. and D.L. Vyas, *Determinants of tractorisation in arid areas of Western Rajasthan*, Agricultural Situation in India, 40(4):257-260, 1985.

Kumar, N. et al., *Farm mechanization and its impact on Punjab agriculture*, paper presented in I.S.A.E. Seminar report, Khalsa College, Amritsar, 1995.

Lekhi, R.K. and Singh, Joginder, *Agricultural Economics*,Kalyani Publishers, New Delhi, July 2002.

Mishra, P. and Sundram, *Some aspects of the economics of harvest combines in Punjab*, Economic & Political Weekly, 10(39), 1975.

PAU, Ludhiana, Directorate of Extension Education, *Package of Practices for Rabi Crops* in Punjab, 2002.

Pathak, A.S., *Farm mechanization and productivity*, Productivity, 23(4): 411-417, 1985.

Planning Commission, Govt. of India, *Expert committee on combine harvesters*, Report (Mimeographed), 1975.

Sethi, K., *Optimum land use pattern and normative demand for tractors in a growing economy-a case study of Ludhiana district*, unpublished M.Sc. Thesis, Punjab Agricultural University, Ludhiana, 1992.

Sharma, B.M.; Goyal, S.K. and G.P. Sharma, "Input use efficiency on tractor and bullock operated farms in Aligarh district of Uttar Pradesh", *Indian Journal of Agricultural Economics*, 47(3): 487, 1992.

Singh, B., *Demand for tractors in Punjab*, unpublished M.Sc. thesis, Punjab Agricultural University, Ludhiana, 1979.

4

Farm Power Sources

Availability of adequate farm power is very crucial for timely farm operations for increasing production and productivity and handling the crop produce to reduce losses. With the increase in cropping intensity the turn around time is drastically reduced and it is not possible to harvest and thresh the standing crop, on one hand, and prepare seed bed and do timely sowing operations of subsequent crop, on the other hand, in the limited time available, unless adequate farm power is available.

Similarly for precision farming, increasing area under irrigation, conservation tillage, straw management and diversification in agriculture, more power is required for water lifting and precision placement/application of agricultural inputs-seed, fertiliser, irrigation water, plant protection chemicals etc and meeting the requirements of diversified agriculture.

The different sources of power available on the farm for doing various mobile and stationary operations are as under:

Sources for Mobile Power

— Human (men, women, children)

— Draught animals (bullocks, buffaloes, camels, horses and ponies, mules and donkeys)

— Tractors

— Power tillers
— Self propelled machines (combines, dozers, reapers, sprayers etc.)

Sources for Stationary Power

— Diesel/oil engines (for pump sets, threshers, sprayers and other stationary operations)
— Electric motors (for pump sets, threshers, sprayers and other stationary operations)

Human Power

The average power availability in sustained working, from a male agricultural worker is considered as 60 watts (0.06 kW) while for a female worker it is considered as 48 watts (0.048 kW) and for child worker as 30 watts (0.030 kW), which are 80% and 50% respectively, compared to male worker. While the population of agricultural workers as percentage of rural population has gone down from about 69.4% in 1951 to about 58.4% in 2001 but in absolute terms, due to increase in overall population, the number of agricultural workers available in rural areas has increased from about 97.2 million in 1951 to about 235.1 million in 2001. The total farm power availability from human power is estimated to be about 1.2×10^5 kW.

These agricultural workers are engaged in different farm operations and depend on agriculture for their livelihood, even when they are not fully employed throughout the year. Due to too much involvement of labour in different farm operations, the cost of production of most of the crops in the country is quite high as compared to developed countries.

Draught Animal Power

Draught animals, particularly bullocks, are still the predominant source of mobile power on about 60% of the cultivated area consisting of about 85 million ha. They are very versatile and dependable source of power and are used in sun and rain under muddy and rough field conditions. They are born and reared in the village system itself and maintained on the feed and fodder available locally. They are ideal for rural transport where proper roads are not available. They reduce dependence on mechanical sources of power and save scarce petroleum products. Their dung and urine are also used as indirect source of energy-farmyard manure, biogas. They also help in maintaining ecological balance.

Under Indian conditions where majority of the people are vegetarian and even amongst non-vegetarians, majority of them don't eat beef, draught animals as by-product of milch animals, will continue to be available for draught purposes in future also. About 4-5 decades back most of the farm operations, water lifting, rural transport, oil extraction, sugarcane crushing, chaff cutting etc, were being done using draught animals only. But with the modernisation of agriculture, development of pucca roads connecting village and availability of electricity in the villages, most of the jobs earlier being done using draught animals, except field operations, are now being done using other convenient and cheaper options.

Over the years the annual use of draught animals is going down. While earlier a pair of animals was being used for about 1200-1800 hours annually, their average annual use has now come down to about 250-500 h only, that too for tillage, sowing, weeding a little bit of rural transport on kuchha roads. The time series population of draught animals during 1971 to 2001 is given in Table 1 which shows that the population of draught animals during 1991-2001 has been going down especially in those states where the demand of tractors and power tillers have gone high. It has been observed that on an average a tractor is replacing about 5 pairs and power tiller about 2 pairs of animals.

The power available from draught animals is related to its body weight. The maximum draft available from different animals, in sustained working, on whole day basis (in two shifts) using local yokes/harnesses have been found as under:

Bullocks : 10-12% of body weight in summer and 12-14% in winter

Buffaloes : 12% of body weight in all seasons

Camels : 18% of body weight up to 7 h, 26% up to 6 h following 2 h work + 2 h rest schedule.

Donkeys : 32% of body weight up to 6 h and 36% up to 4 h in two shifts.

On the basis of the body weight draught animals are categorised as small (200-300 kg), medium (300-400 kg), large (400-500 kg) and heavy (above 500 kg.) animal. From a good pair of animals weighing between 900-1000 kg we can get about 0.75-0.78 kW power. But in most of the States the pair weight of draught animals ranges between 600-800 kg/pair and power availability from them is only about 0.50-0.55 kW/pair.

Power Tillers and Self-Propelled Reapers and Combines

For meeting the increased demand of mobile power for timely farm operations and increased intensity of cropping, additional power is available mainly from tractors and power tillers. Self propelled reapers and combines also provide mobile power specially for harvesting operations. In India tractor and power tiller production started in 1960 and today the annual production has crossed over 250,000 and 17,000 numbers, respectively and India has emerged as number one producer of small tractors in the world. While the demand of tractors has increased steeply, the demand of power tillers has not grown that fast. The demand of tractors that has been increasing steadily has helped in providing additional mobile power on the farm for timely farm operations and has helped in increasing agricultural production and productivity.

There are 14 manufacturers of tractors in the country producing about 60 models of tractors in different hp ranges. About 56.73% of the tractors produced are in the range of 23.12-29.84 kW (31-40 hp), about 17.16% in the range of 30.38-

37.3 kW (41-50 hp) and about 21.50% in the range of 15.66-23.38 kW (21-30 hp). Only about 4.46% tractors are produced in the range of above 37.30 kW (above 50 hp) and 0.15% in the range of less than 14.92 kW (20 hp) range. For calculating power availability from tractors on the farm, a weighted average of 26.1 kW per tractor has been taken.

There are mainly to be checked manufacturers of power tillers in the country producing about 6 models in the range of 5.97-8.95 kW (8-12 hp). In addition to them there are 4 other manufacturers who are importing Chinese make of power tillers and selling in the country. The total sale of power tillers in the country during 2001 was 16018 numbers. The major sales of power tillers are in the States of West Bengal, Tamil Nadu, Karnataka, Assam, Kerala, Orissa and Maharashtra. For calculating power availability from power tillers on the farm, a weighted average of 7.0 kW per unit has been taken.

Stationary Power from Diesel Engines

Stationary power sources in agriculture comprise of diesel engines and electric motors used for irrigation equipment, operating threshers and other stationary machines. For calculation purposes the average weighted power of diesel engines has been taken as 5.6 kW and for electric motors as 3.7 kW and 7.46 kW for small pump sets and submersible pumps respectively.

Farm Power Availability on Indian Farms

Time-series availability of farm power on Indian farms from all sources is given in Table 1.

From the table it can be seen that while in 1951 major contribution in farm power was from animate power (human + draught animal), where as in 2001 the major share was that of mechanical and electrical power. Over the years the contribution of animate source of power, especially that of draught animals, has been going down drastically. This shows that the additional need of farm power is being met through mechanical and electrical sources of power. This trend is going to continue in future also.

Table 1. Farm Power Availability on Indian Farms

Year	Farm Power kW/ha	Source wise%		
		Animate	Mechanical	Electrical
1951	0.25	97.4	2.1	0.5
1961	0.31	94.9	3.7	1.4
1971	0.36 7	9.2	16.3	4.5
1981	0.63	48.2	32.3	19.5
1991	0.92	34.5	34.7	30.8
2001	1.35	18.0	55.0 (36.68)	27.0

Farm Power Availability and Productivity

Considering the average command area of a pair of draught animals, a power tiller of 7.46 kW and a tractor of about 26.1 kW as 2 ha, 6 ha and 15 ha respectively. About 500 pairs of bullocks, or 167 power tillers or about 67 tractors or a combination of the three sources would be required per 1000 ha of cultivated area. Punjab, Haryana, Uttar Pradesh, Bihar and other states do not have adequate farm power to do timely tillage and seedbed preparations. During last 7 years the average sale of tractors was about 74.84% mainly in the northern States of Uttar Pradesh (Including Uttaranchal), Madhya Pradesh (including Chhattisgarh), Punjab, Rajasthan, Haryana, Gujarat and Bihar (including Jharkhand) and only about 19.63% in southern States of Maharashtra, Andhra Pradesh, Karnataka and Tamil Nadu.

This also shows that in future also the southern States (Tamil Nadu, Andhra Pradesh, Karnataka and Kerala), Orissa, West Bengal, Himachal Pradesh, Jammu and Kashmir, Assam and States of NEH region will depend mainly on animate source of power and to a little extent on power tillers. It has been experienced that those states, which have higher horsepower availability on the farm, in general, have higher productivity. This is based on the average farm power availability and food grain productivity of different states in 2001.

The data of AICRP on Energy Requirement in Agriculture Sector shows that the total energy requirement of irrigated farms is much more than unirrigated farms. Similarly the total energy requirements of tractor farms are higher than the mixed farms and animal farms. But since the yields of irrigated farms and that of tractor and mixed farms are higher than the unirrigated and animal farms, the productivity in terms of kg/MJ of energy used is more or less similar.

The grain yield has increased from 612 kg/ha from rainfed farming system to 4677 kg/ha in irrigated farming system using improved equipment. As the energy input through machinery and total energy input increased the grain yield was also increased.

Farm Power to Sustain Agricultural Production

For growing population we need more production of food grains. The ICAR in its Vision 2020 document has projected the demand of food grains as about 293.6 million tonnes by 2020. The Planning Commission of Government of India has estimated a food grain requirement of 230 million tonnes by the end of Xth Five Year Plan (2006-07). Since the cultivated area cannot be increased, the increased production will be possible only by increased productivity and increased cropping intensity. This will call for precision farming and timely farm operations which will require high capacity and precision equipment for which farm power availability will have to be increased substantially. In order to make agricultural production competitive and cost effective the use of mechanical and electrical sources of power will increase in future and the use of draught animals and human power will slowly be going down.

In general, those states, which have higher average horsepower availability per ha on the farm, have high productivity, except states like West Bengal, Tamil Nadu and Kerala where in spite of comparatively lower hp availability, the yields are high. This is because of good and well-distributed rainfall and low water table that reduces the requirement of additional power for irrigation to a great extent.

The additional requirement of food grains in future will be met to a great extent from Indo-Gangetic plains where the potentiality of food grain production is very high. This region will have to be paid greater attention by providing more hp/ha and making availability of agricultural inputs in time. Even now the demand of tractors and other machinery from this region is high and will continue to be so in future also. The productivity of dryland agriculture, which constitutes about 66% of the total cultivated area, is quite low. The productivity of this region has to be increased substantially.

In dryland agriculture, timeliness in farm operations, especially the seedbed preparation and sowing/planting for establishing good crop stand in deficient/receding soil moisture condition is of crucial importance. Unless adequate mobile farm power is not available on the farm, the sowing operation gets delayed resulting poor crop stand and yields. In these areas farmers should use tractors and seed drills/planters for timely sowing operations. Such devices will have to be made available to even marginal and small farmers on custom basis to do their seedbed preparation and sowing operations in time.

Small and marginal categories of farmers constitute 78.2% of total farming households in India. They do not find it economical to maintain even a pair of draught animals. They will slowly switch over to tractor power on custom basis. The demand of tractors, power tillers and matching implements will increase in all the regions, amongst the small and marginal farmers also, on custom basis, which should be promoted by the State and Central governments by giving incentives to entrepreneurs.

At present the availability of farm power in states like Tamil Nadu, Karnataka, Bihar, Assam, Madhya Pradesh, Maharashtra, Rajasthan, Gujarat, Jammu & Kashmir, Jharkhand, Orissa and Chhattisgarh is less than 0.90 kW/ha or even below. These states will continue to use predominantly the draught animal power for few more years before switching to mechanical mobile power. The demand of tractors and power tillers in these states will increase in future. Kerala, which also has low hp/ha availability, will fast be moving towards mechanical power, as the labour wages in this state are the highest in the country.

States in NEH Region, Himachal Pradesh, Jammu & Kashmir and Uttaranchal because of their hilly region, undulating topography and difficult terrain will continue to depend on draught animals and human labour for quite a long time. These states will require lightweight power tillers, which could be transported on hills and operated on terraces.

To meet the demand of food grain production of about 293.6 million tones by 2020 as projected by ICAR, the over all productivity of food grain production at National level will have to be increased from the present level of 1723 kg/ha to about 2300 kg/ha by 2020 for which, besides other things, the average farm power availability will have to be increased from the present level of about 1.35 to 2.00 kW/ha by 2020. Since the draught animal population is going down due to economic considerations and convenience point of view and availability of agricultural labourers will increase only marginally, the additional demand of farm power will come mainly from mechanical and electrical sources of power.

Farm Power in Agriculture

During early sixties, a concept of farm power availability per hectare basis was used to indicate level of farm mechanisation. The potential power availability was used as the measure and could not reflect critical constraint of farm power availability during peak requirements or the actual level of use. The total farm power availability from animate and mechanical sources in 1951-52 was 0.20 kW/ha which increased to 1 kW/ ha in 1996-97. Animate power contributed 60% of the total farm power in 1971-72 and mechanical and electrical together contributed only 40%.

In 1996-97 the contribution from animate power reduced to 21% and from mechanical and electrical power it increased to 79%. The farm power input per unit cultivated land in India is still very low compared to South Korea of 7 kW/ ha, Japan 14 kW/ha and United State of America of 6 kW/ ha. It is evident that mechanical contributed about 78% to the total farm power but for tractive power it is only less than 30% and thus major use of mechanical power has been for stationary farm operations only.

The growth in the ratio of availability of mechanical power to total power availability (per ha basis) and mechanical tractive power to total power availability in Indian agriculture indicates that although 78 per cent of farm power was available from mechanical power sources in 1996, only 29.6 per cent of mechanical tractive power offering higher level of technology was available to Indian farms.

Energy Ratio for Mechanisation Indicator

A major defect in considering power in quantifying mechanisation is that it does not reflect the dimension of time and thus does not bring in the actual scenario in focus. Energy, from this point of view, is a better measure as it can define the actual quantity under use or, in other words, the potentiality of use of available power sources in real terms. Based on use pattern of diesel, electricity, human and animal power in Indian agriculture. It may be seen that the total animate energy availability is decreasing over the years from 2937 MJ/ha in 1971 to 2505 MJ/ha in 1996-97 due to decrease in availability of animal energy.

Human energy use increased from 1,331 to 1,525 MJ/ ha between 1996-97 and 1971. The animal energy use, however, decreased from 1,606 to 980 MJ/ha, indicating a negative annual growth of 1.9 per cent. Increased cost of maintenance of draught animal has been a main cause of this trend. Energy derived from electrical and diesel fuel used by mechanical power sources continued to increase with growth rate of 10.7 and 11.1 per cent respectively to supplement the additional energy need for increased land productivity. The total energy use for crop production increased from 3,282 to 8,773 MJ/ha (annual growth rate of 3.6 per cent) during the period.

The share of mechanical energy over total energy used in crop production has increased from 11 per cent to 76 per cent during the same period. In order to meet the operational energy needs, the contribution of mechanical energy has to be increased substantially. To encourage use of tractors in agriculture, more diesel is required for the purpose. In order to achieve the production targets of food grains from 125 million hectare land, and assuming that the available energy is used for its production, the average specific energy availability on all India basis with present growth rate would be of the order of 5860 MJ/ tonne of food grain as compared to 5480 MJ/ tonne in 1996-97.

Farm Power Sources in Agriculture

The use of farm machinery depends upon the farm power sources available in the country for various tractive and stationary operations. Human and animal power, the two 'renewable energy' or ' bio-energy' sources, have traditionally been used for various farm operations.

The crops are protected from pests, diseases and from weeds, through the application of chemicals. The application of these inputs is achieved through 'human power' in traditional agriculture. Animate power contributed 60.37% of the total farm power in 1971-72 and mechanical and electrical together contributed only 39.63%. In 2001- 02 the contribution from animate power has reduced to 16.38% and mechanical and electrical power, it increased to 83.62% (Table 2).

Table 2. Percentage contribution of different power sources to total power Availability in India

Power source, million kW	1971- 72	1981- 82	1991- 92	2001- 02
Agriculture Workers	15.11	10.92	8.62	6.49
Draught Animals	45.26	27.23	16.55	9.89
Total Animate Power	60.37	38.15	25.17	16.38
Tractors	7.49	19.95	30.21	41.96
Power Tillers	0.26	0.33	0.40	0.54
Diesel Engines	18.11	23.79	23.32	19.86
Electric Motors	13.77	17.78	20.90	21.26
Total Mechanical & Electric Power	39.63	61.85	74.83	83.62
Total Power kW/ha	0.295	0.471	0.759	1.231

This change in the relative contributions of the animate and nonanimate power sources has been due to the requirements of modern agriculture where timely application of crop production inputs is an important factor to maximize return on input investments. The nonanimate power equipment also helps farmers to increase cropping intensities through faster turnaround time between successive crops.

Farm Power Trends

Use of human power

Digging, clod breaking, sowing, interculture, harvesting, threshing, cleaning, and grading are performed by human power using traditional tools and implements. Improved tools have also been developed and commercialised. The agricultural worker population in India increased from 97.2 million in 1951 to 235.1 million in 2001 (Table 3). The agricultural workers comprise of small cultivators and agricultural labourers.

Table 3. Human power in Indian agriculture

Agricultural workers	1951	1961	1971	1981	1991	2001
Cultivators (millions)		99.6	78.3	92.5	110.7	127.6
Agricultural Labourers (millions)		31.5	47.5	55.5	74.6	107.5
Total Human power (million)	97.2	131.1	125.8	148.0	185.3	235.1
Intensity (worker/ha)	0.82	0.98	0.90	1.06	1.30	1.64
Power (Million kW)	4.86	6.55	6.29	7.40	9.26	11.75

Of the total agricultural workers, in 1991, 57.86 million (31%) is comprised of women agricultural workers. The human power available (kW/ ha) has increased from 0.045 in 1971-72 to 0.079 in 2000-01 and the worker intensity has increased from 0.90 to 1.64 worker/ha in the same period. The farm sector therefore, has to absorb more agricultural workers whose work output can be improved by utilising, ergonomically designed tools.

Use of electro-mechanical power

With increased cropping intensity, farmers have supplemented or largely replaced animate power with tractors, power tillers, diesel engines and electric motors. The growth in the electro-mechanical power is reported to be as follows.

Tractive Power: The tractors in India were introduced through importation. There were only 8,635 imported tractors in use in 1951. The local tractor production started in 1961-62 with 880 numbers. Similarly, the manufacture of power tillers started in 1961 with Japanese collaborations. At one time 12 models of power tillers were licensed to be manufactured. However, many of these units closed down or did not even start because of lack of their suitability to Indian farming conditions, poor after-sales service network, etc. presently only two manufacturers are producing power tillers.

Since the pace of production was slow, the Government of India continued to allow limited import of tractors to meet the demand of the farmers till 1974. While approving foreign collaborations, Government of India made it mandatory that tractors to be allowed for manufacturing in India shall be tested under laboratory and field conditions to ensure that they were suitable for Indian farming conditions.

The Central Farm Machinery Training and Testing Institute was mandated to test tractors, power tillers and other farm machinery for the benefit of manufacturers and users. A batch testing scheme was later introduced to enable manufacturers to continuously upgrade the technology and to safeguard user interests. For this purpose, the Government of India fixed norms of specific fuel consumption, noise, vibration, exhaust emission levels, ergonomics and safety measures, and other performance norms. Since then growth in production, quality and performance of tractors and other agricultural equipment has greatly improved.

Sale of Agricultural Tractors

The sale of Agricultural tractors and other farm equipment has increased.

Table 4. Percentage Share of different HP tractors of total sales

HP Size	97-98	98-99	99-00	00-01
<25	7	7	6	4
25-30	12	15	20	20
31-35	51	51	46	45
36-40	9	8	9	10
41-45	5	4	7	8
>45	15	15	12	14

Today more than 250,000 tractors are manufactured every year by 13 manufacturers. These tractors are available in different horsepower ranges of less than 25 to more than 55 horse power. The share of various HP tractors in the recent years has been as follows. Different sizes of tractors are manufactured in India ranging from less than 25 HP to more than 45 HP but most popular range is 31-35 HP (Table 4).

The year-wise production and sale of tractors had shown a healthy trend upto 1999-2000 when it reached a production level of 278556. However, because of drop in sales from 1999-00 due to droughts and floods, and cash flow problems of farmers, the production of tractors dropped in 2001-02 and 2002-03. The production in 2002-03 is reported to have dropped to 168000. Tractor Manufacturers Association (TMA) sources believes that the growth trend shall revive with good monsoon across the country. The Tractor sales show that their demand is regionspecific.

Punjab, Haryana and western UP constituted the major Tractor market in 1980's with 55-57% share of total All India sales. With increased Tractor population in these areas and good successive rains in 1990's, coupled with changes in cropping patterns like adoption of more profitable cash crops (Oilseeds, Pulses, etc.), and better prices, the sales in Gujarat, MP and Rajasthan have been seeing good growth. The contribution of these states improved from 20-22% in early 1990's to 30-32% by the close of the decade. The share of eastern states, namely Bihar, Orissa, West Bengal and Assam had been consistently low at 3-5% due to various socio-economic, agroclimatic and other reasons. The credit availability to the farmers in this area has been another major reason for the slow growth in the eastern states. The tractor sales, since mid 1990's, have increased to about 10-12% of All India Sales.

Tractor sales in Maharashtra, Tamil Nadu, Karnataka and Andhra Pradesh have been showing consistent growth since mid 1980's. Their share in the Indian Tractor industry, which was about 14-15% in 1990, had increased to around 25% in 1997 when the farmers suffered on account of uneven monsoons, poor cotton crops, etc. This region is expected to contribute more than 30% to the tractor industry in this decade. This expectation is based on the fact that the farmers in this southern region have been adopting high value case crops and latest crop production/ management practices.

Production of Power Tillers

The production of power tillers started in 1961 with license to manufacture 12 models. The manufacturers started offering these to framers in various states covering upland and wetland farming conditions. Their introduction coincided with that of agricultural tractors which were more suitable for upland work and provided more comfortable work environment to the operators. The walk-behind power tillers, on the other hand, created dusty environment for the operator. Secondly, the power tillers in dryland conditions were tiresome which resulted in longer rest periods, and consequently affected the work output. These were also difficult to manage in the hilly situations.

The power tiller models being manufactured, and also those being imported from China, etc, and being marketed for wetland, stationary and haulage work are being well received by the farmers. The 7 available models have a Drawbar power between 5.3 kW to 10.7 kW. Their production and sale has also increased to about 16000 units annually. Their population has reached 122000 in 2000-01, which is expected to cross 180000 in 2005-06.

Sources of Stationary Power

Electric Motors and Diesel Engines are the primary sources of stationary power for irrigation, threshing and various post-harvest agro-processing operations. Diesel Engine population, which was 1.443 million in 1971-72 increased to 5.528 in 1995-96, and, is expected to cross 7.4 million in 2005-06. Electric Motor population has increased from 1.535 million in 1971-72 to 7.464 million in 1995-96, and, is projected to be around 12 million by 2005-06. The above shows a substantially faster growth of electric motors which is due to their higher efficiency, lower maintenance hassles and spread of rural electrification coupled with preferential power tariffs to farmers. The studies on operational efficiency of irrigation pumps have shown the efficiency of electric motor operated pumps to be 31.1% against only 12.7% of diesel engine operated pumps.

References

Anonymous, *Production and sale of tractors,* Agricultural Engineering Today Vol 26(1-2), 2002.

De, Dipankar, *Research on Energy Requirement in Agricultural Sector under AICRP on ERAS,* Agricultural Engineering Today Vol. 22(3-4): 60-85, 1998.

Singh, G., *Data book on Mechanisation and Agro-Processing since Independence,* Technical Bulletin, CIAE Bhopal, 1997.

Srivastava, N.S.L., *Role of Agricultural Engineering in Doubling Food Production in next ten years,* Agricultural Engineering Today, Vol. 23(1-2): 37-49, 1999.

5

Impact of Mechanisation on Agricultural Productivity

Farm mechanisation is regarded as sine-qua-non to reduce the human drudgery and enhance the agricultural productivity. During the post-green revolution period, the impact of farm mechanisation on agricultural production and productivity had been well recognised in India. Depending upon the use of other inputs such as irrigation, high yielding seed varieties, chemical fertilisers, herbicides and pesticides, different States in India have attained different levels of mechanisation. Consequently the agricultural production & productivity has witnessed three to four fold increase. Studies have been conducted by various organisations & individuals which have highlighted the impact of agricultural mechanisation on farm production and productivity.

Singh and Chancellor found that though, tractor and tubewell farms had significantly higher yields than bullock farms in case of wheat, much of the difference was accounted for by difference in other factors such as level of irrigation etc. The use of tubewell was found to be associated with significantly higher yields compared to the persian wheel irrigation. ITES, Madras found that tractor—owned farms obtained increased productivity of paddy, sugarcane and groundnut by 4.1 to 28.3 %, 13.1 to 34.2 % and 9.8 to 54.8 % with an average value of 15.8 %, 23.2 % and 31.8 % respectively. Likewise, the average increase of productivity on farms hiring tractors was reported to be 11.8 %, 13.0 % and 16.0 % for paddy, sugarcane and groundnut respectively.

Pathak et al conducted survey on five different categories of farms in Ludhiana District of Punjab to assess the effect of power sources on production and productivity. The yield of paddy, maize and wheat was reported to be higher on tractor farms than on bullock farms.

The yield of wheat after paddy or maize was significantly higher on tractor farms than bullock farms. The use of tractors enhanced agricultural productivity due to better seedbed preparation, timeliness of operations and precision in distribution and placement of seed and fertiliser owing to the use of the seed-cum-fertilizer drills.

NCAER conducted a survey of farms owning tractors, using tractors on custom-hire and owning bullocks in seven States belonging to three major agroclimatic zones. A sample of 815 farming households was selected randomly from 85 villages. It was reported that an average tractor-owning farm obtained higher yields than a bullock farm, which varied from crop to crop and ranged from 72 % in the case of sorghum to 7 % in the case of cotton. Tractor users also obtained higher yields compared to bullock farms.

Aggarwal analysed data of a sample of 240 farms situated in the principal wheat growing areas of Punjab for the crop year 1971-72. The study revealed that the use of tractors instead of bullocks for ploughing and sowing did not add to the yield of high yielding varieties of wheat. It appeared that the advantage of timeliness of operation of a tractor was nominal, when the crop preceding to wheat could be harvested and threshed sufficiently in advance and the land could be released for timely wheat ploughing and sowing even on bullock farms. Unlike the neutral effect of the tractor use on wheat yield, the use of tubewells in comparison to canals was found to have a significant positive effect on the productivity.

Nandal and Rai conducted a study by dividing Haryana in three homogenous zones on the basis of intensity of mechanisation. In all, 54 farms were selected from each of the three zones making a total sample of 162 farming households. The impact of mechanisation on crop yield was studied on three different categories of farms.

It was apparent from the study that the tractor-operated farms had higher yield of wheat and paddy. In case of farms using tractors on custom—hire basis, the yield was comparatively low. The study revealed that tractor-owing farms invariably used higher level of agricultural inputs and had better control on timeliness of operations.

Balishter, Gupta and Singh conducted a study in Mathura district of Uttar Pradesh on the basis of three levels of mechanisation, i.e.

- — non-mechanised farms having neither tubewell nor tractor
- — partially mechanised farms having only tubewell, and
- — mechanised farms having both tubewell and tractor.

The yield was reported to be higher by about 10 to 27 % in mechanised farms and by about 2 to 26 % in partially mechanised farms in comparison with nonmechanised farms for all the major crops grown on the sample farms.

Higher yields on tractorised farms were associated with higher levels of fertilizer and irrigation use, but without statistical testing. An exception was the study conducted in Punjab by Kahlon which found no statistically significant yield effects.

In the two other areas, the effect did not apply to all farms. No significant differences for high yielding rice varieties were found. Differences were significant for maize in one area and cotton in another.

According to Motilal in Delhi Territory, yields increased significantly for paddy (13.7%), wheat (15.9%), and sugarcane (29.7%). However, tractor farms used 35% more fertilizer, so these increases could not be entirely, attributed to tractors. In Uttar Pradesh, tractors gave a yield advantage of 17.6% in sugarcane and 41% in wheat, but significance tests and fertilizer inputs were not reported.

In the NCAER study, yield increases with tractors accompanied increased fertilizer use, but sample sizes were small. In other areas, large yield increases were reported for summer paddy in Bihar (28.6%), for desi paddy in kharif and high yielding paddy in Andhra Pradesh, and for groundnuts in Coimbatore (23.9%). However, fertilizer applications were also higher (Bihar, 31.8%; Andhra Pradesh, 36.3%; Coimbatore, 28.7%).

Finally, Singh and Chancellor's regression analysis on 26 maize farms in Meerut district showed no significant effect of tractorisation on productivity.

Rao conducted a study to investigate the effect of the use of tractors on yield, employment of labour and cropping pattern. The methodology employed was on recall basis i.e. data obtained before and after tractor acquisition.

As many as 4000 tractor-owning farmers were contacted through a mailed questionnaire but only 1500 responded. The study revealed that of all the crops raised on different sizes of farms, tractor owning farms obtained higher yields per ha and the increase was more for the larger sizes of tractor owning farms.

Binswanger adopted value of crop production per ha per year as a proxy to determine any cropping effects attributable to tractorisation. Production value was considered as a function of cropping intensity, yields, cropping patterns, and a residual interactive effect.

Since tractorisation had increased the first two only slightly, it was reasonable to suppose that the increase in production value was attributable to a tractor-based cropping pattern or to factors such as irrigation, credit, and/or managerial capacity. Laxaminarayan et al reported 6% grain loss in harvesting and threshing with traditional methods, and 2-4% with combines. A multiple regression analysis indicated that greater fertiliser use mainly explained combine users' higher yields.

Singh and Chandra used the linear regression function to examine the effect of important inputs on crop productivity for the State of Punjab. Standardised regression co-efficients were calculated for relative efficiency of different inputs.

The elasticity of productivity for fertilizer, irrigation and farm power was reported to be significant in the production function. Relative efficiency of farm power was higher followed by fertilizer and irrigation.

The co-efficient for relative efficiency and standardised regression coefficient for fertilizer was reported to be 0.23, irrigation, 0.35 and farm power 0.45. The effect of farm power, however, showed decreasing effect beyond 3.24 kW/ha. The best fit mathematical relationship for growth of food grain yield and use of fertiliser could be expressed, with statistically significant value of R^2 = 0.96 at 1 % level by:

$$y = 1160.4 + 13.972X_f \; ; \; (R^2 = 0.96)$$

Where, y = state average food grain yield, kg/ha and Xf = state average fertilizer consumption, kg/ha The best fit mathematical relationship between yield and % area irrigated was reported to be as follows:

$$y = 278.32 \; e^{0.0262 \; X_i} \; ; \; (R^2 = 0.98)$$

Where Xi = average food grain area under irrigation,%. The value of coefficient of determination (R^2 =0.98) was reported to be the maximum and significant at 1 percent level. Best mathematical relationship between yield and unit power could be represented by quadratic function and was given as,

$$y = 534.41 + 1924.9X_p - 297.53 \; X_p^2 \; ; \; (R^2 = 0.99)$$

Where, Xp = state average availability of power, kW/ha. The co-efficient of determination (R^2 = 0.99) was statistically significant at 1 % level. Singh suggested an analytical approach to study the growth dynamics of agricultural inputs and their effect on productivity in Madhya Pradesh. He found that the impact of farm power upto 1kW/ha was not significant.

Standardised regression coefficients estimated values based on time series data, exhibited that fertilizer (0.42) was the most important input followed by irrigation (0.38) and power (0.18) but from the spatial data for the year 1996-1997 it emerged that irrigation (0.52) played the most vital role followed by fertiliser (0.38) and power (0.05) in increasing the crop productivity.

Input-grain yield relationship computed on the basis of equations of time series and spatial data also revealed identical trends, and therefore either of the regression equation could be used for planning input requirements for higher yield.

Effect of Mechanisation on Cropping Intensity

Agricultural mechanisation has made significant contribution in enhancing cropping intensity. The growth in irrigated areas and tractor density has had direct bearing on the cropping intensity. Findings of the studies conducted in the past are briefly presented to highlight the contribution of mechani-sation in enhancing the cropping intensity.

Chopra carried out a study on a sample of Punjab farms. He made a comparison of tractor-owning farms in terms of the situation before and after the introduction of tractors. The cropping intensity was reported to be higher after the introduction of tractors.

NCAER conducted a study of tractorised and non-tractorised farms in nine States of India. The study revealed that tractor-owning farms had a higher cropping intensity of 137.5 % as compared to 131.8 % in the case of those without a tractor.

Cropping intensity was found to be generally higher on small farms. Among the States surveyed, cropping intensity of tractorised farms was the highest in the Punjab, followed by Uttar Pradesh, Tamil Nadu and the lowest in Rajasthan and Maharashtra.

In a study by Pathak et al., on a sample of 115 farms in Ludhiana district, the average cropping intensity with fodder crop was reported to be higher on bullock farms than on tractor farms. However the cropping intensity without fodder crop was comparatively higher for tractor farms. On medium size farms of 6 to 12 hectares, the cropping intensity with fodder crop on tractor farms was 180 % as compared to 174.5 % for the bullock farms. The cropping intensity without fodder crop on tractor farms was 153.9 percent as compared to 149.6 percent on bullock farms.

The cropping intensity with fodder was higher on bullock farms than on tractor farms whereas the cropping intensity without fodder was lower. But when comparing the cropping intensities in the two categories having medium size of holdings (6 to 12 hectares), the tractor farms were in a better position since the cropping intensity with and without fodder on these farms was 3.2 and 2.8% higher than on bullock farms respectively. The higher intensity was attributed to the availability of more mobile power on tractor farms than on bullock farms.

The differences were lower and statistically not significant. But it was observed that bullocks became the major constraint and thus the bullock farms began to depend on the use of tractors through custom-hiring. The group of the bullock farms feared that if tractors were not made available to them, the output and cropping intensity would decline. NCAER conducted a comprehensive study to ascertain the factors motivating a farmer to own or use/ hire a tractor and also the impact of tractorisation on rural income and agricultural employment.

A sample of 815 farming households from 85 villages of 7 States belonging to three major agroclimatic zones was selected. To study the impact of tractor use on cropping intensity, the land use of tractorised farms was compared with that of the bullock farms.

The cropping intensity of tractor-owning and tractor-using households in all the States was higher than that of bullock farms. An average tractor owner had a cropping intensity of 12 % higher than that of the bullock farms. Similarly, the cropping intensity of tractor using households exceeded that of bullock farms by 8 %. Cropping intensity of small farms was found to be uniformly higher than that of larger farms and those with the facility of a tractor had raised their cropping intensities further (Table 1).

Table 1. Cropping Intensity of seven selected states by Farm Size (Percent)

Farm size (ha)	Tractor-owner	Tractor-user	Bullock-farm
Below 2	180	172	161
2-4	176	170	151
4-10	163	154	148
Above 10	168	140	148
All Farms	165	156	149

Aggarwal reported that the use of tractors and tubewells, in comparison with the use of bullocks and canals had a significant bearing on the cropping intensity. However, the advantage of tubewells over canals was found much greater than that of tractors over bullocks.

Among the tractor-using farms, it was essentially those owning tractors which accounted for higher average cropping intensity. However, those depending solely on hired tractors were not found to differ much from the bullock farms in regard to cropping intensity. As the farm size increased, the cropping intensity declined. The percent gross cropped area irrigated was positively related to the cropping intensity.

The study conducted by Nandal and Rai concluded that the cropping intensity showed consistently positive relationship with tractorisation. Within the given size groups, tractor-owning farms had higher cropping intensity as compared to tractor-using or animal operated farms.

Patil and Sirohi conducted a study to measure the impact of tractorisation on the irrigated farms of Ahmednagar District of Maharashtra State. In Maharashtra, the duration of sugarcane crop varied from one to one and half years depending on the planting season. Thus, cropping intensity

(I) and cropping intensity (II) was calculated considering sugarcane as a single crop and as equivalent to two crops, respectively. The cropping intensity on small sized TOF (Tractor operated farm) and TOF + BOF (Bullock operated farm) was significantly higher in comparison with the small sizes BOF and BOF+THF (Tractor hired farm), when the cropping intensity pattern (I) was considered. But the differences in cropping intensity on medium and large-sized farms were not significant among different categories of farms.

The cropping intensity on all size groups of farms of TOF and TOF+BOF was also significantly higher in comparison with the BOF and BOF+THF when the cropping pattern (II) was considered. This was attributed to allocation of higher proportion of area for cultivation of longer duration sugarcane crop on tractor-owning farms.

Balishter et al. based on their study in Mathura district of Uttar Pradesh, concluded that the average cropping intensity was the highest in mechanised farms having tubewell and tractor (206.4%) followed by partially mechanised farms having only tubewell (176.6%) and nonmechanised farms (143.8%). Thus, facilities of assured tubewell irrigation and mechanical power helped the farmers in raising the cropping intensity on their farms significantly.

Singh argued that, cropping intensity was mainly dependent on annual water availability. Rainfall distribution in a year as well as the percent area irrigated in different seasons reflected the water availability. The State of Punjab had low annual rainfall of 555 mm concentrated mainly during three months (July, August, and September) but with 94 % area irrigated had a cropping intensity of 1.80.

The average annual yield of rice in India was about 2 t/ha. However, Punjab had already achieved an annual yield of 5.3 t/ha. If the whole of India could achieve this level of annual yield, a grain production, of over 500 million tons, enough to feed a population of 1.7 billion could be achieved. However, the yields obtained in many countries for a single crop were as high as 7 t/ ha for rice in Korea, 7.6 t/ha for wheat in France and U.K. and 9.3 t/ha for maize in Italy. Thus, even Punjab had a potential for further enhancement of the crop productivity.

Effect of Mechanisation on Employment of Human Labour

The impact of farm mechanisation on labour employment, particularly in a labour surplus country like India, has been a matter of concern and debate. The available evidences suggest that mechanisation had helped in overall increase in employment of human labour. A study by Rao and Singh on "Tractorisation in Kanjhawala Block in Delhi Territory" had showed that both tractor as well as non-tractor farms had on an average 8.2 persons per farm and the labour force at their disposal was neither surplus nor inadequate.

GIPE, Poona concluded that tractorisation generated greater demand for labour by facilitating more intensive cultivation. Thus, there was no significant displacement of human labour after tractorisation.

UPAU reported that mechanisation accompanied by use of new seed technology and adoption of modern cultivation methods had a beneficial effect on employment. Kahlon, reported that reduction in aggregate labour use on tractoroperated farms owning tubewells was only 1.3% as compared to bullock operated farms.

Billing and Singh studied the changes in the demand for labour. The aggregate impact of adopting the improved technologies in Punjab resulted in reduction of human labour employment to the tune of 11.5% compared to conventional level. In Maharashtra, the reduction was negligible at 0.2%.

Bhagwati Committee on Unemployment concluded that mechanisation of agricultural operations, by and large, displaced bullock labour and not human labour. In another study, Johl reported that increased use of tractors was associated with marked rise in employment due to their effect on cropping intensity. AERC, Delhi conducted a study on a sample of 96 farms in Haryana. It was concluded that the use of tractors had, in most cases, displaced only one pair of bullocks. The overall human labour input for crop production per cultivated hectare was practically the same for both types of farms. USAID, New Delhi conducted a study to assess the impact of tractorisation.

The large scale adoption of high yielding varieties accompanied by higher use of chemical fertilisers and enhanced cropping intensity led to higher demand for farm labour.

Permanent labour on tractor holdings showed a decline to the tune of 2.2% compared to the conventional level. The employment of casual male labour showed an increase of 38.5%. In respect of labour employment, the increase was even higher at 80%. A tractor on an average displaced four bullocks.

Babu conducted a study in West Godavari district of Andhra Pradesh and reported that tractor use induced changes in cropping pattern which promoted the demand for human labour.

Grewal reported that the employment of bullock labour on mechanised farms was 85 % less than those, which used only bullock labour for farm operations. The average reduction in labour input on mechanised farms was 11.9 % in comparison to nonmechanised farms. The average output per man-days on tractoroperated farms was about 42 % higher than on bullock operated farms.

According to Harrington restraint on the rate of agricultural mechanisation was necessary to solve the problem of underemployment in India. In his opinion, more labour would be employed for land preparation, water management, weed control, insect and disease control, harvesting and threshing when higher yields were obtained by adopting HYV technology.

Singh and Singh conducted a study in Muzaffarnagar district of Western U.P. and observed that mechanisation did not displace labour significantly. The average human labour employment per hectare on mechanised farms was 169 man-days as compared to 180 man-days on nonmechanised farms, which was only 6.1 % less. It was concluded that displacement of human labour was significant only on mechanised farms in the size group of 15-20 ha and above.

NCAER studied the impact of mechanisation on employment in Muzaffarnagar district of Uttar Pradesh. It was concluded that the labour input per hectare increased as one switched over from nonmechanised farms to farms using tubewells in both small and medium size groups. In small tubewell farms, the increase was much larger over nonmechanised farms. When a thresher in addition to a tractor and tubewell was acquired, there was additional labour input per hectare for crop cultivation in both small and medium sized farms as compared to tractor plus tubewell farms.

The mechanised small farms used 5.7 % more labour. Labour displacement was observed on medium size farms during initial stages of mechanisation. As the level of mechanisation increased, there was a significant gain in labour employment. It was not possible to assess the labour input for several non-farm activities directly related to the use of tractors and farm equipment. If this was also taken into account, the labour input for each level of mechanisation might go up further correspon-dingly. AERC, Viswa Bharati conducted a study on a sample of 60 farms in Shahabad District of Bihar. It revealed that the displacement of human labour in terms of value of all the inputs was 11.6% in case of farms owning tractors and 5.1% in case of farms using tractors.

The maintenance of tractors by the first category of farms led to additional employment of labour, which more or less compensated for displacement of labour in farming operations. Animal power was completely replaced by tractor power by the farms in the first category, while the farms in the second category continued to make partial use of animal power combining it with tractor-hire services.

NCAER conducted a survey of tractorised and non-tractorised farms in 9 States to study the impact of tractorisation on employment, economics of tractorisation, horsepower preferences and subsidiary uses of tractors.

In the case of paddy, tractor using farms showed a 3 % higher use of human labour compared to farms using animal labour. A tractor replaced the bovine stock to the extent of 50 %. IIM, Ahmedabad concluded that tractorisation did not lead to decrease in human labour employment.

On the contrary, it helped in relaxing the constraints of the operational area which increased employment on farms significantly. In fact, tractor farms employed more casual and permanent labour.

Another study by IIM covered a sample of 161 farms drawn from three major crop zones of Gujarat State. The study revealed that tractor farms utilised less human labour as compared to custom farms and bullock farms. As the level of mechanisation increased, use of bullocks steadily decreased. However, upto a certain intermediate level of tractor use, only bullock power was displaced. Beyond this stage, a marginal reduction in manual labour was also observed.

IEG, Delhi analysed the economics of technological changes such as the use of high yielding varieties of seeds, tractorisation etc. in Indian agriculture. The study covered the country as a whole with main focus on Punjab and deltaic Andhra Pradesh. The study reported that the technological displacement of labour associated with tractor use was compensated by the employment of labour owing to increased yield as a result of tractor use among farms characterised by partial tractorisation. The net employment effect of tractor use turned out to be positive when its complementarity with other techniques was taken into account.

Rao considered the effect of tractor use on employment by comparing farms, which owned tractors with those that did not. He concluded that the tractors had neutral effect on farm employment. Based on a study conducted in Tiruchirapalli district of Tamil Nadu, AERC, Madras (1975-76) reported that the human labour input accounted for 32 % for bullock farms while it accounted for 22 to 23 % for the tractor-owing and tractor hiring farms.

ITES, Madras studied the displacement of labour due to mechanisation of agriculture on a total sample of 510 farmers under three levels of mechanisation in seven districts of Tamil Nadu. The study covered the principal crops of the State and revealed that the displacement of labour due to mechanisation was more than offset by the employment created through increased production, additional area brought under cultivation and higher intensity of cropping.

An additional area of about 425 ha was brought under cultivation in 170 farms selected for the study with an average value of 2.5 ha per farm. Additional human labour employment varied from 96.22 to 559.52 man-days giving an average value of 330.57 man-days per farm.

Singh and Singh in their study on "Impact of Farm Mechanisation on Human and Bullock Labour use" in two regions of U.P. reported that for seed bed preparation, one hour use of tractor replaced 5.12 labour-days whereas the corresponding figures for medium and large farms were 1.61 and 1.29 labour days respectively.

Likewise, the magnitude of decrease in labour employment due to an additional one hour use of tractor in land preparation decreased with an increase in farm size in Meerut District. An additional hour of tractor would replace 1.04, 0.76 and 0.13 human and bullock labour days on small, medium and large farms respectively.

Thus, the rate of labour displacement through tractorisation was higher on relatively small farms of Jaunpur District than large farms of Meerut District of U.P. This was obvious because the negative relation between cropping intensity and farm size was steeper for bullock farms than for tractor farms. In a comparative study of tractorised and bullock operated farms in Purnea district of Bihar, Singh and

Goswami reported that the human labour days employed on an average cropped hectare on tractorised farm worked out to 87.6 man-days with 113.9 man-days on custom-hiring farms and 120.6 man-days on bullock operated farms which were reduced to 3.10 % on tractorised farms. The percentage share of labour requirement for interculture, irrigation, harvesting and threshing operation was higher in case of tractorised farms as compared to bullock operated farms.

NCAER studied the implications of tractorisation on employment adopting a cross-section comparison of tractorised households with bullock-operated farms. In conventional practice, the bullocks provided the motive power and implements were guided by human labour.

Certain amount of human labour displacement was inevitable in the operations where a tractor was employed. However, tractorised farms produced more output on a hectare of land by following more scientific agronomic practices and applying optimal doses of material inputs and thereby provided employment to more labour.

Tractorised farms also raised commercial crops on a relatively larger area. Such changes in the cropping pattern had a favourable impact on the total employment. An average tractor owner employed 110.5 man-days per cropped hectare as compared to 104.8 man-days by bullock farms. Thus, the aggregate employment of tractor owners was 5.4 % higher than that provided by bullock farms. Tractor-using farms as a group employed 0.8 % additional man-days on a cropped hectare as compared to bullock farms. The survey revealed that tractorised farms employed less number of persons in operations where a tractor was used, but more in other operations. This helped in increasing the aggregate employment.

Aggarwal concluded that the tractors replaced mainly family labour time on small farms and permanent labour time on large ones. Use of threshers displaced mainly family and casual labour time on small farms and family and permanent labour time on large ones. The combined effect of family labour time was increased in the use of permanent and casual labour time on farms of most size groups. With the addition of power threshers, these effects were lessened.

Patil and Sirohi studied the employment per hectare of cropped area. The total labour employment was the highest on small farms and decreased as the farm size increased in respect of all categories of farms. The overall human labour employment was the highest for tractor operated farms followed by tractor plus bullock operated farms. The ratio of family labour to total labour employed per hectare decreased with an increase in farm size.

On an average, the per hectare employment of hired labour and total labour was higher by 39 % and 24 % on tractor operated farms and by 43 % and 22 % on tractor plus bullock operated farms respectively than that of bullock operated farms. The higher percentage of hired labour employment with the increase in farm size, in general, and of tractor-owning farms in particular disproved the general opinion held regarding displacement of human labour by mechanised farming.

Sidhu and Grewal analysed human labour employment in Punjab based on the data collected for the agricultural year 1981-82 under tractor operated farms and tractor operated plus combine harvester using farms. The study showed that tractorisation did not replace human labour. There was no significant difference in human labour use on tractor operated farms and bullock operated farms. The introduction of harvester combines, however, significantly reduced the use of human labour on the farms.

Balishter, Gupta and Singh reported human labour utilisation based on their study conducted in Mathura district of Uttar Pradesh for the crop year 1984-85. It was reported that in absolute terms, there was no decrease in human labour employment due to mechanisation. In fact, there was an increase which could be attributed to increase in the cropping intensity and shift in cropping pattern from less labour intensive to more labour intensive crops. It was also reported that with mechanisation, the demand for hired labour increased while participation of family labour in crop production declined. Thus, mechanisation reduced the use of family labour while it increased the use of hired labour.

Almost all the studies which reported reduction in human labour employment resulting from mechanisation attributed it largely to loss of human labour associated with maintenance of bullock labour hours, which got reduced as a result of mechanisation. But the more important reason seemed to be wrong inferences drawn from unadjusted farm size, and a comparison of labour use made on all sizes of bullock operated farms with that of large tractor operated farms. Thus, in the presence of mixed effects of farm size and all other variables on employment, it was difficult to lean heavily on the findings of these studies.

Grewal and Kahlon attempted to control such variables as irrigation and farm size at the sampling stage. Their study showed that in the situation of matching size farms, the labour use on tractor farms was higher than that of bullock farms in case of groundnut, cotton and potato regions.

Desai and Gopinath in their study on "Impact of Farm Tractorization on Productivity and Employment (Gujarat State)" reported that the effect of tractor use was positive on employment. According to this study, with an increase of every hectare in area, the employment went up by 29.43 and 38.74 man-days on tractor and custom hiring farms. The study showed that for every hundred rupees increase in output, employment increased by 3 days on tractor and 2 days on custom and bullock farms, thereby showing a close relationship between output and employment.

Thus, the positive impact of rapid growth in output on employment could offset the effect of displacement of human labour by tractor power. These results rejected the findings of Binswanger who observed that tractor farms had much less labour input per unit of output and measured in this way, labour displacement could be very large. No doubt, a technical change could reduce demand for factors including labour per unit of output, but the expansion in production would be more than enough to offset this reduction, which means the rapid growth in output would bring about not only increase in productivity but also increase in employment, as had happened in the Punjab situation.

Rao argued that 'tractors would not lead to unemployment in regions of expanding agricultural output because they were in response to genuine shortage and high cost of draft power. In so far as tractors saved the resources allocated for maintaining draft animals and enabled an increase in output by raising cropping intensity and yield per hectare, their use proved socially beneficial in the context of the shortage and high prices of agricultural commodities.

Kahlon studied the "Impact of Mechanisation on Punjab Agriculture with special reference to Tractorisation" under a Bank study. The pattern of employment of human labour per holding showed that family labour was employed more on the pure bullock farms, causal labour was hired more on the pure tractor farms and permanent labour was employed more on the tractor with bullock farms. It was found that family labour increased by 17.65 % with shift from pure bullock to tractor plus bullock technology and total labour employment per cropped hectare declined by just 5.64 %.

Assuming simultaneous nature of relations in the employment of family labour, hired labour and permanent labour, the demand function was fitted for various categories of labour, using two stage least squares technique. The study showed that as for the geometric means for other factors affecting labour demand, the total labour input increased very slightly on pure tractor farms, 512.12 man hours per hectare, in comparison with 504.97 man hours on pure bullock farms, with very little change in the composition on the labour force.

Kahlon's study showed that total labour employment per cropped hectare did not fall by any significant amount on tractorised farms except on very large tractor farms of an average size of 23.28 hectares which had further declined in number owing to laws of inheritance and ceilings on holdings. There was enough evidence to show that the large bullock operated farms, when mechanised, were able to obtain a cropping intensity and employment that matched well with small and medium bullock operated farms and this gain from the introduction of the tractors could not be minimised. Studies made by Kahlon, Grewal, Verma and Sharma supported these results.

Verma reported that on small and medium size farms, the increase in the requirement of labour due to increased cropping intensity was not enough to compensate for the decrease in the requirement of labour due to mechanisation. However, on large farms, higher cropping intensity increased the requirement of labour more than the labour displaced due to mechanisation.

Binswanger interprets this to say 'that in all cases where there was substantial increase in labour use by tractor farms, it was associated with shifts in cropping pattern or irrigation which were an outgrowth of the better overall capital availability rather than that of the tractors per hectare'. One wonders if all this development would come about in the absence of mechanisation. If not, why separate out these effects through mechanistic statistical tests. It would not be out of place to mention that most of the mechanisation studies were conducted during the early years of the introduction of the tractors i.e. during 60's and 70's. The opposition to this technology and particularly to the introduction of tractors was based on the findings of some of these studies, which showed that the differences in timeliness of operations and productivity per hectare did not come out to be statistically significant on tractor farms as compared with bullock farms.

While quoting such findings, the limitations of inadequate infrastructure for machinery maintenance and repair and the time it took to master the mechanical skills, which largely contributed to these results, were conveniently forgotten. Against this background, it would be wrong to conclude that tractorisation did not improve the timeliness and output of farm operations or increased secondary and tertiary employment only marginally.

Subsidiary and Non-farm Employment

Different studies conducted on farm mechanisation indicated that net human labour displacement in agricultural operations was not significant and it was more than compensated by increased demand for human labour due to multiple cropping, greater intensity of cultivation and higher yields.

Table 2. Indirect Employment generated by Farm Mechanisation in Selected Districts

Sr. No.	District	No. of Persons employed
1	Chingelpet	403
2	South Arcot	473
3	North Arcot	439
4	Tiruchirapalli	467
5	Thanjavur	695
6	Coimbatore	528
7	Madurai	646
	Total	3651

On the other hand, the demand for non-farm labour for manufacture, services, distribution, repair and maintenance as well as other complementary functions increased substantially and helped in relieving rural unemployment to some extent. Mechanisation in agriculture provided indirect employment to skilled and unskilled persons engaged in operation, repair and maintenance of prime movers and farm equipment.

ITES, Madras quantified the number of persons employed in maintaining the tractors and operating them in the field. Table 2 gives the extent of indirect employment generated in the government and private organisations at the district level including private workshops.

Mechanisation in agriculture in seven selected districts had provided employment to 3651 persons per year. Agricultural households undertook subsidiary farm activities including dairying, poultry-keeping etc. The survey conducted by NCAER (1980) provided details of employment created through such activities and taken up by different households as shown in Table 3.

Table 3. Subsidiary Employment per Hectare (Man-days)

	Tractor-owner	Tractor-user	Bullock-farm
Family labour	17.6	33.7	48.6
Hired labour	11.6	10.1	7.0
Total labour	29.2	43.8	55.6

Though, the total subsidiary employment of tractor owners and users was less than that of bullock farms, they employed more hired labour. Thus, demand for hired labour in case of tractorised farms had increased both in case of crop production and subsidiary activities. The non-farm employment generated by mechanisation provided employment through manufacture of farm machinery, distribution of the spare parts, fuel and lubricants and repair and maintenance of tractors, engines, and other machines. Thus, the marginal loss in direct farm employment resulting from farm mechanisation was more than offset by non-farm employment in agro-industrial activities and secondary and tertiary sectors of the economy.

NCAER assessed the employment in the manufacture, distribution, repair and servicing of tractors based on sample survey of private repair workshops and manufacturer's authorised service establishments. It was concluded that a private repair shop, in general, generated employment of 22.2 man-days per tractor per year.

According to Baig, there was labour displacement in the area of ploughing and transport which was, however, more than compensated by higher employment in other agricultural operations notably, fertilising, weeding, interculture, pest control, irrigation, harvesting, threshing etc. Tractors also generated indirect employment in providing skilled and semiskilled work in distribution, maintenance, service etc. The estimated employment pattern is summarised as below:

Indirect Employment due to use of Tractors

Employment in the organised sector

— Employment in 12 operating tractor plants 15,000

— Employment in the auto-ancillary industries 28% of their total employment of 1,60,000 workers. 45,000 Total 60,000

Employment in rural based small industries

— Manufactures of Implements, Trailers and accessories:

- — 20% value of 40,00 new tractors 36 crores of which 40% labour-14.4 crores @ 200 p.m. 60,000
- — Purchase by 3,50,000 old tractors @ 2000/- p.a.- 70 crores of which 40% labour -28 crores support workers @ 200 p.m. 1,17,000

— Sales, service & repairs

- — 1,200 authorised dealer points 12,000
- — Repairs of 3,50,000 tractors @ 2,500 p.a.-87.5 crores of which 40% labour -35.5 crores support @ 200/- per worker 97,000

— Sales of diesel & lubricants (10% of total of POL outlets) 14,000 Total 3,00,000

Effect of Mechanisation on Nonfarm Income

Mechanisation has generated many non-farming and subsidiary activities among the farming households. On one hand additional employment was created in the manufacture of farm machinery, distribution of the equipment and spare parts, fuel and lubricants, repair and servicing etc., on the other hand many subsidiary activities like dairying and poultry keeping got generated. NCAER showed that tractorised farms reduced their draught animal stock and increased their milch stock. Tractor-owners and tractor-users had 82% and 25% more milch cattles, respectively as compared to bullock farms.

A tractor owner was able to increase his household income by undertaking supplementary activities such as dairying and provisions of custom-hiring. A tractorowner with a land holdings of 6.28 ha, had an average gross income of Rs. 47,534 which exceeded that of a bullock farm and tractor-user household by 285% and 132%, respectively.

Effect of Mechanisation on Gross Farm Income and Net Return

Farm mechanisation has greatly helped the farming community in the overall economic upliftment. The studies conducted on impact of mechanisation on farm income clearly support this view -point. AERC conducted a series of studies related to economics of mechanisation. These studies revealed that the gross income was higher on mechanised farms than nonmechanised farms.

The gross crop output per cultivated hectare was reported to be Rs.3144 for tractor operated farms as compared to value Rs.2677 for bullock operated farms NCAER reported that the tractor farms secured 21% more income per hectare of gross cultivated area compared to bullock farms.

The net return per hectare of gross cropped area or net cultivated area was higher for tractorised farms than the non-tractorised farms as a result of better utilisation of resources. Another study by NCAER revealed that the tractor owners and users derived higher per hectare gross income compared to bullock farms.

The gross income per hectare of an average tractor-owned house hold was 63% higher than that of a household using only bullock labour. The gross income per hectare of tractor-using households as a group exceeded that of the bullock farms by 31%. The average net return from a tractor-owning farm on a cropped hectare exceeded that of a bullock farm by 152%.

A tractor using farm also derived a net additional income of 84% over a bullock farm. A tractor-owning farm spent 57% more than bullock on material inputs and 62% more on human labour. An average tractor owner and user, inspite of spending more on cultivation expenses, derived higher net income on a cropped hectare compared to bullock farm. However, this should not be attributed entirely to tractor usage as other factors such as hybrid seeds, fertiliser and irrigation also contributed to it. In the study in an irrigated area of Ahmednagar district of Maharashtra, Patil and Sirohi reported that, on an average, the gross return were higher by about 33 to 34% on tractor-owning farms than those on bullock operated farm.

References

Asian Productivity Organization, *Farm Mechanization in Asia*, APO, Tokyo, 1983.

Baig, M.A., *The Tractor in India - A significant instrument for future development*, Proc. of the Symposiumon Farm Mechanization - Problems and Prospects ISAE, North Chapter and ISAE, PAU, Ludhiana pp. 57-66, 1978.

Balishter, Gupta, V.K. and Singh, R., *Impact of Mechanization on Employment and Farm Productivity*, Productivity, 32 (3): 484-489, 1991.

Chopra, K., *Tractorization and Its Impact on the Agrarian Economy of Punjab*, A Report submitted to USAID, American Embassy, New Delhi, 1972.

Desai, D.K & Gopinath, C., *Impact of Farm Tractorization on Productivity and Employment (Gujarat State) Centre for Management in Agriculture*, IIM, Ahmedabad, 1975.

6

Agricultural Machinery Industry

Blacksmiths and carpenters have been the traditional fabricators of agricultural equipment in India. The early agricultural mechanisation in India was greatly influenced by the technological development in England. In 1889, Watts and Kaisar introduced ploughs, corn grinders and chaff cutters. Sardar Joginder Singh, who was the Agriculture Minister in the Punjab Governments, introduced the steam tractors in India in 1914 for reclamation of waste land and eradication of 'Kans' a type of perenial grass.

Horse drawn and steam tractor operated implements were imported during the latter part of the 19th century. The horse drawn equipment imported from England were not suitable for bullocks and he-buffaloes used in India and thus, were suitably modified by small scale manufacturers to suit Indian draught animals. With the establishment of Allahabad Agricultural Institute, Allahabad in 1942, the development activities in agricultural machinery accelerated and as a result bullock drawn Meston, Shabash and Wah-Wah ploughs were introduced in Uttar Pradesh, manufactured by the Agricultural Development Society, Naini (Allahabad) in early forties.

The Indian farmers gradually responded to farm mechanisation technology especially after Green revolution in 1960s. High yielding varieties with assured irrigation and higher rate of application of fertiliser gave higher yields and better economic returns. This enabled the farmers to start adopting mechanisation. The development of power thresher with integrated Bhusa making attachment and aspirator blower and mechanical sieves for grain and straw separation in 1960s was the major achievement of Indian engineers which was widely adopted by our farmers. Gradually demand for other farm machinery such as reaper and combine harvester also increased.

Demand of tractors in the country was met through importation until 1961 when Eicher Tractors Ltd. and Tractors and Farm Equipment Ltd started manufacturing tractors with foreign collaborations. To meet the additional demand, importation continued up to 1977. Meanwhile many other industries started manufacturing tractors with foreign know how such as Gujarat Tractors Ltd, Escorts Ltd, International Tractors (India) Ltd., and Hindustan Machine Tools Ltd. Punjab Tractors Ltd. started their production with indigenous technology in 1974. Many more industries started manufacturing tractors since then with indigenous and foreign know how.

Improved Machinery

Preparation of seed bed

Deshi ploughs, bakhar and patela were the most popular traditional implements for seed bed preparation prior to 1960's. Cultivator, disc harrow, mould board plough, puddler, disc harrow-cum-puddler, peg tooth harrow, spring tine harrow, rotavator and patela harrow etc. operated by animal and tractor are the improved implements which have been adopted by farmers. The growth in use of tractor drawn machinery has been in the range of 9-17%. Different sizes of cultivators and disc harrows are used but due to farm road and terrain constraints, cultivators of more than 15 tines and disc harrows of more than 18 discs are not much in use. The power from higher horse power tractors, therefore, is not fully utilised.

Sowing and planting equipment

The line sowing not only saves seed but also facilitates regulated application of fertiliser near root zone. Besides, it helps control of weeds through use of mechanical weeders. The animal drawn Dufan (two row), Tifan (three row), Enatigoru and FESPO plough (all local sowing devices) are used by the farmers as these cover more area and cost less. For precise application of seed and fertiliser, mechanically metered seed drills and seed-cum-fertiliser drills operated by human, animals and tractors have been developed and are being manufactured to suit specific crops and regions.

Interculture and plant protection equipment

Weed control in irrigated and rain-fed agriculture during Kharif is a serious problem and the yield is affected to the extent of 20-60%, if not controlled. Khurpi is the most popular tool used for removal of weeds but it takes 300-700 man-hours to cover one hectare.

Table 1. Trends in growth of harvesting and threshing machinery

(in hundred)

Power source	1971-72	1991-92
Tractor Powered Combine Harvesters	3.5	61.5
Self Propelled Combine Harvesters	4.5	35.0
Stationary Threshers	2058	13793
Wheat Threshers	1825	10757
Paddy Threshers	136	1353
Others Threshers	97	1683

Use of long handle wheel hoe and peg type weeders, reduce weeding time to 25-110 hours. Bullock operated weeder and cultivator are also used for control of weeds. Different designs of low cost hand operated sprayers and dusters are available for plant protection. Spraying in cotton, paddy, sugarcanes, fruits and vegetables, oilseeds and pulses has become popular.

Harvesting and threshing

The technology for development of harvesting and threshing equipment is motivated by following factors:

- Economic considerations-reduction in cost of production and reduction in harvest and post harvest losses and quality of produce,
- Social realities- non-availability of labour during the harvesting period and to ensure timeliness, and
- Ergonomic considerations -reducing drudgery in the operations.

The harvesting systems prevalent in the country include (Table 1):

- Harvesting with sickle followed by threshing with animal trampling.
- Harvesting with sickle and manual threshing.
- Harvesting with sickle/reaper and threshing with stationary power threshers. Use of reaper harvester is limited due to collection, bundling and transport cost of material and by-products.
- Stationary power threshers varying from 5 to 15 hp, operated by diesel engines and electric motors, and tractor pto power. The present trend is to use high capacity machines on custom basis.
- Combine harvesting.

Sickles are widely used for harvesting. These are easily available at low cost in the villages but their output is low. Self-sharpening serrated sickles of better material and design have been developed. Sickles would continue to be used for various crops by small and marginal farmers, and in hilly regions. Reapers powered by engines, power tillers and tractors have been developed and introduced for harvesting wheat, paddy, soybean, ragi and mustard. However, because of labour and cost involved in the collection, bundling and transport, their adoption is expected to be limited.

Traditionally, threshing of wheat and barley was being done by bullock trampling which is arduous and time consuming. The mechanical threshers of varying power range (5-15 hp) are commercially being manufactured which not only thresh the grain but also provide good quality Bhusa. Farmers use self-owned threshers or on custom hiring basis. More than 70-80% wheat, barley, gram, soybean, sorghum and pearl millet crops are estimated to be threshed by mechanical power threshers. Paddy crop is easy to thresh by beating but losses are quite high.

Pedal operated paddy threshers reduce drudgery. These have become popular in Eastern India. Raspbar type paddy threshers cause less breakage to paddy stem and, thus, straw can be put to better use. These threshers have become popular in Andhra Pradesh, Tamil Nadu, Karnataka and Kerala. In regions where work force availability is inadequate, harvesting with combine harvesters is in vogue.

Combine harvesters are being preferred as they reduce the turn-around time to facilitate increased cropping intensity. Tractor-powered and self-propelled combine harvesters are being manufactured in India. About 700-800 combines are sold annually. Track-type Combine harvesters, especially suitable for paddy crop, are also being manufactured locally. The combine harvesting of wheat, paddy and soybean has been well accepted by farmers.

Apart from the work force availability problems, the usage of combined harvesters has helped in timely harvesting of grains avoiding losses due to adverse weather conditions. The Tractor Powered Combine Harvesters, costing only 25-30% of the self-propelled combines, may be owned by individual farmers.

The self-propelled combines are largely owned by custom-hiring contractors. The demand of combine harvesters for harvesting wheat crop in Punjab, Haryana and Uttar Pradesh encouraged the local manufacturers to develop local combines. More than 48 manufacturers, mainly in

Punjab, produce self-propelled and tractor operated combines for harvesting wheat, paddy, soybean and gram.

The entrepreneurs from Rajasthan and Punjab provide combine harvesting services to the farmers in the states of Maharashtra, Gujarat, Madhya Pradesh, Andhra Pradesh and Tamil Nadu. Most of the combines are suitable for harvesting wheat and paddy. Farmers also use them for gram and soybean.

Post Harvest Equipment

Agro-processing includes farm-level processing to improve quality of produce and technology for loss prevention in storage, handling and transport. Major equipment which have been developed and adapted for farm level processing include cleaners, graders, dryers, shellers, decorticators, storage structures milling equipment etc. Cottage and industrial level secondary processing includes, rice mills, grain mills, dal mills, oil mills, preservation and processing of animals, fruits and vegetables etc. to increase shelf life and their quality. The growth of these machines is given in Table 2.

Table 2. Status of post harvest equipment

Equipment	Number in 1991	Projection for 2000
Cleaners and graders	110,000	290,000
Dryers	7,000	25,000
Flour units	266,000	350,000
Dal mills	10,000	25,000
Rice mills	130,000	150,000
Oil expellers	222,000	450,000
Maize shellers	65,000	115,000

Today more than 73% of paddy, 55% maize, 24% pulses and 45% oilseeds and 45% sugarcanes are processed by modern machinery besides other commercial crops.

The total turn over of food market is approximately Rs. 250000 crores as estimated by Ministry of Food Processing of which value added products comprised of Rs 80000 crores.

The total export is estimated at about Rs 11000 crores with rice contributing 29% and marine products 42%. The agriculture processing sector has immense employment potential for rural people, provided these activities are under taken largely in rural areas.

Adoption of Machinery by Size of Holdings

It is generally believed that only farmers having large acreage have adopted mechanisation inputs. However, based on data from Input Survey 1981-82, 1986-87 and 1992 of Department of Agriculture and Cooperation, Ministry of Agriculture Government of India, reveals that farmers of all farm holding sizes have adopted implements and equipment from bullock-drawn implements to pump sets, threshers and tractors. Infact, the data shows that the percentage growth of tractors, pump sets and threshers has been higher in the marginal and small farm holding categories.

Manufacture of Agricultural Machinery in India

The adoption of mechanisation technology depends upon the local manufacture and after-sales-services besides credit and financial incentive provided by the Government.

The manufacture of agricultural machinery in India is quite complex comprising from village artisans, tiny units, small scale industries to State Agro-Industrial Development Corporations and organised tractor, engine and processing equipment industries. Traditional hand tools and bullock drawn implements are largely fabricated by village craftsmen and small scale industries.

Organised sectors manufacture sophisticated machinery such as tractors, engines, mills and dairying equipment. The small-scale industries seldom have R&D facilities and they depend upon public institutions for technological support. They require not only drawings but also prototypes and technical guidance to manufacture the equipment. These industries however, upgrade the technology with experience.

Classification of Industries

The classification of industries in India is based on total capital investment (plant and machinery) rather number of workers employed. These are

- — village craftsmen,
- — cottage industries,
- — tiny industries,
- — small scale industries,
- — medium scale industries and
- — large scale industries.

This classification was done to help the small-scale units through incentives and marketing support.

Table 3. Classification of industries in India

Category Maximum	capital investment
Village artisans and cottage industries	Unorganized in rural areas
Tiny industries	Rs. 15 lakhs
Small scale industries	Rs. 1 crore
Medium scale industries	Rs. 5 crores

The limit of investment is given in Table 3.

Village craftsmen

Village artisans are the main source of supply and repair and maintenance of hand tools and traditional implements are made by village craftsmen. These include implements and tools like khurpi, spade, sickle, local ploughs, bakhar, sowing devices, yokes, patela, leveller, oil ghanis, grinding wheels, hand mills, handoperated milk churning tools, winnowing devices, sieves, wooden storage structures, bullock carts, manual water lifting devices etc. If village artisans are properly trained they will accelerate the adoption of mechanisation inputs due to their proximity with farmers.

Tiny and small-scale industries

The tiny and small scale units fabricate bulk of improved agricultural machinery such as ploughs, cultivators, disc ploughs and harrows, seed drills, planters, plant protection equipment, reaper harvesters, combine harvesters, threshers, cleaners, graders, mills, crushers, oil expellers, diesel engines, irrigation pumps, dairy machinery etc.

Agricultural machines are reserved for small-scale units. There are more than 18000 such units scattered all over the country but have concentration in selected regions. Some of these units also fabricate implements and equipment for tractor and power tiller manufacturers. They may lack good machine tools and heat treatment facilities. Some of them are more organised and have better fabrication toolings and thus are able to manufacture better quality machinery.

The bulk of the farm machinery is made by the small-scale industries. They use materials from mild steel to medium carbon steel. Heat treatment practices are generally inadequate except in few industries manufacturing knife & tillage tools.

Equipment manufactured by the SSI units includes Soil working tools, seeding & planting equipment, hand hoes, sprayers & dusters, harvesting & threshing equipment, like reapers, threshers, combines, maize shellers, decorticators, cleaners, graders, mills, oil expellers etc.

Organised farm machinery industries

The medium scale and large scale industries manufacture diesel engines, electric motors, irrigation pumps, sprayers and dusters, land development machinery, tractors, power tillers, post harvest and processing machinery and dairy equipment. There are 13 tractor, 2 power tiller, 200 diesel engine, 600 irrigation pump, 48 combine and 188 earthmoving machinery manufacturers.

The marketing of agricultural machinery by these industries is through their network of dealerships and, therefore, these manufacturers are able to provide effective after-sales service. These industries upgrade their product and process technologies through their own R&D efforts, in addition to technological support from external agencies.

Standardisation and Quality

The Bureau of Indian Standards (BIS) with its network of centers and laboratories in the country is mandated to ensure quality manufacture and marketing of agricultural and industrial products in the country.

The BIS formulates specifications of agricultural machinery and other appliances and prescribes test codes. The Bureau also issues ISI quality certificate marks to the products which meet technical specification as per BIS standards. The list of standards released by BIS is given in Table 4.

Table 4. List of standards on agricultural machinery

Type of machinery	Standards Nos.
1. Tractors and power tillers and engines	160
2. Soil working equipment	47
3. Sowing and fertilizer application	28
4. Irrigation & drainage equipment and system	30
5. Crop protection	27
6. Harvesting and threshing	24
7. Horticulture and plantation	55
8. Processing machinery	25
9. Milling equipment	18
10. Dairy and animal husbandry equipment	58
11. Farm transport	14
12. Storage structures	53

The Government of India has established Farm Machinery Training & Testing Centres for promotion of quality farm machinery. In case of agricultural machinery, requirement of quality certification is limited to the sale of agricultural machinery financed under Government Schemes.

On certain items connected with safety and health hazards, it is mandatory to have minimum safety standards built into the design or in the installation of machinery during operation. Threshers, chaff cutters, sugarcane crushers etc. fall under this category. The agricultural machines manufactured by the organised sector like tractors, earth moving machinery, irrigation equipment, plant protection, dairy equipment, processing machinery etc. are certified for their quality by BIS.

The adoption of agricultural machinery is greatly influenced by the quality and after sales available to the farmers. Since manufacture of agricultural machinery is reserved for small-scale industries, the quality is affected by the manufacturing technology adopted by them.

Testing and evaluation helps in up-gradation and quality production of machinery. R&D institutions and quality certification agencies conduct the T&E. Testing and evaluation is conducted on newly developed equipment and during its serial production, to facilitate and ensure quality, reliability, durability, functional ease, comfort in operation and cost of operation. Testing is conducted with well defined standard parameters, defined in BIS, ISO, or OECD standards, and where as evaluation is done to measure the performance under simulated or field conditions for the parameters for which the equipment has been designed.

The Bureau of Indian Standards has the statutory authority to inspect the quality of products manufactured and marketed in India. The agricultural machines manufactured by the organised sector like tractors, earth moving machinery, irrigation equipment, plant protection, dairy equipment, processing machinery etc. are certified for their quality by BIS. The BIS has established their Regional Testing Laboratories to facilitate testing and evaluation, including that of agricultural machinery. The BIS has also authorised other Government and Semi-Government testing laboratories to conduct testing on their behalf as per BIS Test Codes or ISO Test Codes.

The Ministry of Agriculture, Government of India has established 6 Regional Testing Centres located at Budni, Madhya Pradesh (Central Region), Hissar, Haryana (Northern Region), Ganganagar, Rajasthan (Northern), Assam (Eastern Region), Anantpur, Andhra Pradesh (Southern Region), Tamil Nadu (Southern Region). CFMT&TI, Budni is equipped to undertake testing of tractors, combines and other agricultural machinery. Other Centres, test agricultural machinery and irrigation equipment. These Centres conduct testing and evaluation as per BIS Test Codes.

The Regional Research Laboratories (RRL) under CSIR have established microprocessor based modern testing facilities especially for metallographic and material testing. These laboratories have memorandum of understanding (MOU) for sharing research and testing facilities for quality assessment of products, including agricultural machinery.

Few NGOs, institutions/association have established independent institutions for undertaking testing and evaluation and these have been authorised by Govt. Deptts./ BIS/other consumer organisations (Bank Consumers Protection Forum, etc.) to undertake quality performance testing on their behalf. Few Universities, including Agricultural Universities and Institutes under Indian Council of Agricultural Research conduct quality certification evaluation for consumers on request.

Agricultural Machinery Marketing

The large and medium scale manufacturers have well organised distributors and dealers through out the country to undertake advertising and product promotion in their respective territories, conduct product awareness training programmes for the prospective customers, provide after-sales-service to the customers including free services, repair and maintenance, supply of parts, etc. Therefore, this organised sector has the whole of the country as their market due to which their production volumes are large, and their information feed back about their product performance, improvements required in design, production processing or quality, and the new requirements of the farmers to undertake product developments.

Very few small-scale industries have established their marketing network and therefore provide service support in their premises. In the absence of standardisation of parts and components farmers are compelled to carry their machines to the manufactures for repair and replacement of parts and components. Due to this, their market size is limited to their proximity, and they are not able to develop their businesses. The village artisans on the other hand are located in the villages and therefore provide immediate attention to the needs of the farmers in their immediate neighbourhoods. Therefore, the tools and implements, etc. made by them are against specific requirements of individual customers.

Agricultural Equipment Development

The ICAR is the apex institution for promotion of education, research and extension in the agricultural sector. Besides, Department of Science and Technology and Council for Scientific and Industrial Research (CSIR) also promote development of technology for mechanisation and agro-processing. National Research Development Corporation and Poly-technology Transfer Center under CSIR market technology developed in the country. The District Industries Centers are the nodal department for promotion of village industries and small scale industries.

Khadi and village Industries Commission (KVIC) promote and market products produced by cottage industries. The Indian Council of Agricultural Research for the first time sponsored a scheme to conduct state-wise survey of existing tools and implements used by the farmers in 1954. The results were published in the form of a book entitled "Indigenous Agricultural Implements of India" in 1960s. During the sixties, the ICAR made serious efforts to promote research and development on improved farm implements by establishing 17 Research Training and Testing Centres (RTTCs), one in each major state, which were operated by the State Departments of Agriculture. The major mandate of these RTTCs was to test and modify existing implements and develop new improved implements suitable for different agroclimatic conditions of the country.

During the latter part of the sixties (IV Five Year Plan Period) two Zonal Research and Testing Centres, one at IARI, New Delhi and the other at TNAU, Coimbatore and four research centres at Ludhiana, Pune, Hyderabad and Mandi were established to promote use of improved machinery. Besides, All India Coordinated Research Projects (AICRPs) have been established all over the country under ICAR research system to undertake regional research needs and conduct front-line demonstrations of technology with the cooperation of State Agricultural Universities. The ICAR also sponsored All India Coordinated Research Project (AICRP) on Energy Requirement in Agriculture in 1970 to study the role of improved machinery in conservation of energy and increasing productivity.

The ICAR also sponsored an AICRP on Research & Development of Farm Implements & Machinery, Production of Prototypes and their Evaluation under Different Agrocli-matic conditions in 1972 to develop and promote suitable implements for different agroclimatic conditions. The ICAR established the Central Institute of Agricultural Engineering at Bhopal in 1976 and Central Institute of Post Harvest Engineering and Technology at Ludhiana in 1989 to support R&D, technology transfer and prototype manufacturing activities. The research in the area of farm machinery was further strengthened with the creation of other AICRPs on Power Tiller, Human Engineering and Safety in Agriculture and a NRC on Reducing Drudgery of Women in Agriculture.

The ICAR has initiated Front Line Demonstrations of Improved Machinery for piloting the technology with the assistance of Department of Agricultural Cooperation of Ministry of Agriculture, Government of India. Receptive farmers are provided implements, service and training to create awareness. Prototype workshops have been established for developing commercial grade machines for pilot introduction.

The ICAR through AICRPs on agricultural machinery conducts demonstrations on farmers field under front line demonstrations (FLD) on newly developed machinery. The ICAR has established 261 Farm Science Centers (Krishi Vigyan Kendras). The proposal is to have at least one KVK in each of the 500 districts. These KVKs are mandated for verification of the technology in addition to providing skill oriented training to the farmers and village artisans.

The assimilation of R & D requires an effective technological infrastructure of institutions and services to develop and test prototypes, to set up pilot plants for intensive evaluation and extensive demonstrations besides, training and credit support. New technology also requires network for transfer of technology to the manufactures. Popularisation of agricultural machinery in the country is undertaken by the Provincial Governments through Department of Agriculture or Department of Agricultural Engineering.

The activities are coordinated by the Department of Agriculture in Cooperation with the Ministry of Agriculture, Government of India. The Ministry of Food Processing promotes technology related to agro-processing. The extension system deals with the first-line extension projects with a view to:

— demonstrating the latest technologies to the farmers as well as the extension agencies;
— testing and verifying the technologies on the farmers field

— providing opportunities to get firsthand scientific feedback;
— developing extension or technological models for the state extension systems;
— providing training and communication support; and
— promoting research in transfer of technologies.

Central Sector Extension Programmes

— *Promotion of agricultural mechanization:* For this programme a plan outlay of Rs. 340 million had been made during VIII plan. In this scheme, subsidy for the purchase of tractors below 18 pto hp along with 3 matching implements was provided to farmers, individually or in a group, having irrigated land between 2.4 to 3.2 ha. Subsidy rate was 30% of the cost to a maximum limit of Rs. 30000. This programme has been slightly modified during 1996-97 by enhancing the plan outlay from Rs. 340 million to 450 million and enhancing the scope of tractors to be included from 18 pto hp to fuel efficient tractors fitted with engines not exceeding 1800 cc (tractors in the range of 15-30 hp). With the revised outlay of Rs. 450 million, it would cover subsidy for 15,000 tractors and matching implements.

— *Special food production programmes:* Under these programmes subsidy is provided up to 50% of the cost limited to Rs. 1500/- per implement/ farm on bullock drawn implements. Under maize and millet programmes, subsidy is also provided on plant protection equipment limited to 50% of cost or up to Rs.600/-

— *Oil Seed Production Programme:* Under this programme subsidy is being provided on bullock drawn implements to the tune of 50% limited to a maximum of Rs. 700/- to small and marginal farmers. Subsidy is also being provided on plant protection equipment up to 50% of cost limited to Rs. 500/-

— *National Pulse Development Programme:* Under this programme subsidy is being provided on bullock drawn implements up to 50% of cost limited to Rs.500/-per farmer. Subsidy is also provided on plant protection equipment up to 50% of cost limited to Rs. 500 and Dal processing equipment up to 50% of cost limited to Rs. 4000 for machines of less than 1.5 hp capacity.

— *Intensive Cotton Development Programme:* Under this programme subsidy is being provided on plant protection equipment up to 50% of cost limited to Rs.600 for manually operated equipment and 25% of cost limited to Rs. 4000 for tractor mounted equipment.

— *Integrated Programme for Rice Development:* Under this programme subsidy is being provided on animal drawn implements, rice transplanter and water pumps up to 50% of cost limited to Rs.1500. Subsidy is also being provided on power tillers up to 25% of cost limited to Rs.12000.

— *Development of Industrial Designs of Prototypes of Implements:* This scheme was approved for the 8th plan with an outlay of Rs.13.5 million. The scheme envisages identification of improved, nearly developed equipment and grant of financial assistance to the R&D institutions for developing industrial designs together with jigs and fixtures needed for their commercialisation.

State Sector Extension Programmes

Under the state sector programmes generally two promotional programmes are being perused by the state governments;

— *Custom Hiring of Implements:* Under this programmes tractors with matching implements for deep ploughing, rotavation, reapers, threshers, combines, drilling and boring machines, dozers, ditchers etc are being given to farmers on subsidised hire charges.

— *Manufacture of Implements:* State governments having agricultural implement workshops are manufacturing a good number of improved agricultural implements and providing to farmers at nominal profits. Most of the implement popularisation Schemes sponsored by the Central Government are under review to have an integrated approach to the input requirements of the farmers. Under Macro Subsidy Schemes, the State Governments are required to prepare a consolidated requirement based on the inputs to be promoted for increasing the productivity of agriculture.

Credit and Financial Incentives

The purchasing power of the farmers is low. The government provides subsidy and credit at reduced rate to the farmers who are economically and socially at disadvantageous position to adopt modern technologies. The long-term credit are usually availed for the purchase of mechanisation inputs and short term for the purchase of seed, fertiliser etc. This is one of the indicators of progressive attitude of farmers. The agricultural machines and tractors are purchased through credit, available from organised financial institutions.

NABARD is the main refinancing institution. The Government also provides incentives to farmers for modernisation of agriculture. This is linked to crop specific programmes operated by state governments. Some of the states could not avail the advantages in the absence of adequate infrastructure for promotion of agricultural engineering programmes. The state government may have to strengthen their extension machinery for providing incentives to the farmers. Draft Agricultural Policy resolution emphasized special consideration for input support to poor farmers with fragmented land holdings and those in eastern, hilly regions, rainfed and drought prone areas. The financial requirement for the purchase of agricultural machinery has increased considerably. The commercial and Cooperative Banks provide credit for the purchase of machinery.

The total long-term institutional credit provided to agriculture sector which is usually for non-recurring expenditure during 1998-99 was Rs.132080 million. Of this, Rs.43920 million was provided from Cooperative Banks and Rs.88160 million from Commercial Banks. 1998-99. The medium and long-term loans are usually disbursed for the purchase of machinery. To simplify the credit and other financial incentives "Farmers Agri-Credit Card" facility is being introduced in few states to facilitate easy availability of credit including fiscal incentives. An estimated Rs 119580 million is required based on 1997-98 prices for the purchase of major farm machinery. This estimate does not include budgetary requirement for sprinkler and drip and processing machinery. Farmers prefer to invest their own money for the purchase of hand tools and bullock drawn machinery and for others they avail the credits from the Banks.

Future Agricultural Mechanisation Technology

India is a large country with wide-agro ecological diversity having predominance of rainfed agriculture, with irrigated agriculture limited to 34% only. Farm holdings are small due to higher population density and land fragmentation will continue due to 'Laws of Inheritance' and 'Hindu Succession Act'. Majority of the farmers have limited surplus money to modernise farms or to invest in improved inputs. Draught animals and increasing agricultural workers population may remain to be the major source of farm power for soil manipulation and for crop handling, particularly in Hill and Mountain regions.

Mechanical power for tillage, irrigation, harvesting and threshing will be preferred, including on custom hiring basis. As a result of GATT agreement, prospects of agro-export are likely to increase and product quality standards stipulated under WTO would encourage more and more farmers to adopt modern agricultural production technologies. The future agricultural mechanisation technology package therefore may have to;

— be eco-friendly utilising land water and bioresource catering to the varied group of farm holders,

— facilitate farming operations which are arduous and hazardous,

— increase productivity and conserve resources through effective utilisation of chemical, biological and mechanical inputs, and

— modernise commercial agriculture to facilitate agro-export.

The mechanisation policy may have to be distinctly different to serve hill agriculture, low lying water logged soils, rainfed and irrigated lands and regions having agro export potential.

Hand tools for handling of soil, improved sickle, weeder, sprayer, duster, sheller, decorticator, etc are being manufactured by unorganised and organised sector and these are being used within the means and resources of the farmers. The quality of these equipment will further improve as demand picks up and general quality consciousness increases amongst farmers. Tools for horticulture and forestry will have growing demand. The present gap in their availability could attract manufacturers to import these. This would prompt manufacture of better design and better quality tools in India.

Use of mechanical power is increasing but draught animal power will continue to be in use for many years to come. These draught animals will continue to be utilised for tillage, sowing, interculture and transport. The present available designs of farm machines for these operations are adequate but will require refinement in quality through material substitution and better manufacturing processes. For the manufacture of critical parts and components in large volumes, the role of organised sector should be encouraged to adopt the village craftsmen and small scale industries in order to help them in the assembly and production of good quality agricultural machinery.

Tractors, Power Tillers, Diesel Engines

Tractors

The present tractor industry is capable of meeting domestic requirements with 10-12% annual growth leading to stabilisation at around 7-8%. The export of tractors, showing growth trends, is going to increase particularly in Africa, and many Asian and CIS countries. The tractor manufacturers are aware of these possibilities and some of them have already taken steps to make their products more export worthy through their own R&D and joint ventures.

This will not only help export but also Indian farmers. The latest package of computer integrated flexible manufacturing system will not only be economical but also provide better industrial environment. The present tractor use is limited to tillage, transport and stationary operations and, thus has a lower annual utilisation. In future, use of tractors will extend from primary to secondary operations like sowing, spraying, interculture, harvesting, agro-forestry, tree harvesting, plantation, land development, excavation for drainage, mulching, drilling etc. These will require introduction of specialised tractors.

Power tillers

Demand has not picked up in the country due to availability alternative power sources. For mechanisation of hill agriculture and orchards, light weight power tillers with matching equipment for different farm operations would be required. The present designs may not suit this requirement and, therefore, more suitable designs, including imported machines, should be introduced with adequate incentives and credit support to popularise their use.

Irrigation equipment

Demand for irrigation equipment including drip and sprinkler is increasing, particularly in water deficit regions. The drip system is likely to increase for application of chemicals and fertilisers. The pumps are operated by electric motors or diesel engines. While quality of electric motors is satisfactory, the farmers continue to use horisontal and vertical type diesel engine operated pumps. These engines, though cheap, have higher operational costs due to poor quality and life, and high fuel consumption. Efficient lightweight diesel engines have to be developed, manufactured and promoted. Solar photovoltaic pumps and wind mill pumps have potential especially for drip irrigation systems.

Power operated agricultural machinery

Machines for primary and secondary tillage operations, of varying quality, are commercially available to meet the farmers needs. In the absence of quality promotion measures, the quality of these machines is rather poor. A policy should be devised to promote the manufacture of these machines conforming to BIS specifications, and their sales subject to BIS quality certification.

Till such time, the manufacturers have to be encouraged to manufacture these machines by adopting BIS certified components. Several designs of sowing and planting machinery have been developed and commercialised. The utility of these machines in rainfed agriculture has been amply demonstrated. The R&D Institutions and BIS should work together to screen the designs for their standardisation to promote manufacture of good quality implement.

Transplanting of paddy, sugarcane, vegetables and trees are yet to be developed to an acceptable level before these are taken up for commercial production and adopted by farmers. This should be taken up on priority by R&D Institutions and industries. Initial importation may accelerate the pace of development.

Equipment for application of manure and liquid fertilisers are not available in India. Handling and application of biogas slurry is also manual. These require special attention of R & D Institutions.

Plant protection equipment

Plant protection manufacturing industry is fairly well organised. Hand and power operated machines are manufactured by large number of industries some of which have obtained ISI certifications. Machines for tall crops and trees including ULV and electrostatic sprayers are yet to be taken up for manufactured by the industries. The development of horticulture sector is likely to increase the demand of such sprayers.

Harvesting and other harvesting equipment

Combine harvesting is in vogue in some States and the industry is capable of meeting the present demand. The combine harvesting at present is limited to cereal crops only. However, most of the straw is left in the fields and burned. Alternative straw handling and disposal technology may have to be developed and promoted as burning of straw is creating environmental pollution and farmers are loosing valuable animal feed material.

Maize, sorghum, cotton, sugarcane, potato, peanut, sunflower, safflower, soybean and pulses are predominantly harvested manually. Large and commercial farming of these crops would require alternative harvesting and handling machinery. Initially demand could be met through limited importation which will lead to local adoption and manufacture. Similarly, harvesting of fruits has to be mechanised for timely and damage free harvesting. Their use will also reduce danger and drudgery involved in fruit harvesting. A suitable strategy and programme should be put together for selection of appropriate designs, their field evaluation, design optimisation and manufacturing in India.

Threshing technology

Threshing technology is well accepted by the farmers. The demand for cheep threshers led to design of thresher with inadequate safety measures resulting in to a large number of fatal accidents. The Government enacted Safety Laws; the

Dangerous Machine (Regulation) Act 1983. Since, the agricultural machines are manufactured by unorganised sector, the enforcement of the Act has been difficult. A policy decision should be taken to make it mandatory to display the source of their origin for ensuring adherence of minimum safety standards by the manufacturers.

Machines for land development, excavation for drainage channels, mulching, trench cutting and post hole digging are not commonly available. Fodder production, tree felling, pruning etc. are performed manually. In the absence of machines for such operations, even an assessment of their requirement is difficult. Limited introduction through importation may help in projecting their need and the likely future demand. Green house technology is increasingly becoming popular in some States for horticulture and floriculture. Implements and equipment is required for the management of crop production in green houses.

References

Government of India, *Indian Agriculture in Brief*, 24th Edition, Directorate of Economics and Statistics Department of Agriculture & Cooperation, Government of India, 1992.

_____________, 1994 &1995. Agricultural Statistics at a Glance, Directorate of Economics and Statistics, Department of Agriculture & Cooperation, Government of India.

Input Survey 1981-82 and 1986-87, *Ministry of Agriculture*, Government of India.

Singh, G., *Agricultural Mechanization and Agro- Processing after Independence*, Central Institute of Agricultural Engineering, Bhopal, 1997.

_____________ and Bharadwaj, K.C., *Directory of Agricultural Machinery and Manufacturers*, Central Institute of Agricultural Engineering, Bhopal, 1985.

_____________, Gyanendra., Characteristic and use of draught animal power in India, Indian Journal of Animal Sciences 69 (8): 621-627, 1999.

7

Custom Hiring of Agricultural Machinery

The composition of Indian farms varies drastically. India has a large number of small farms with land holding of less than 2 hectares and very poor economic condition. Single farm ownership and use of tractors and machinery on these small farms is not economically viable. But through custom hiring of agricultural machinery even small farmers have been able to get the benefit of agricultural mechanisation. Punjab and Haryana have amply demonstrated it.

A sea change has occurred in Indian agriculture since Independence. Pressure on land has increased. Progressive mechanisation and other developments such as establishment of agroindustries and agro-service centers, rural credit facilities, etc.

Composition of Farms

The percentage of marginal and small farms in India was 69.6, 74.9 and 78.2 in 1970-71, 1980-81 and 1990-91 respectively (Table 1). Marginal and small farms dominate in Indian agriculture in spite of the fact that Consolidation of Land Holding and Fragmentation Act, 1948 was enacted long back. The situation in case of Punjab is different. Percentage of marginal and small holdings actually decreased between 1970-71 and 1980-81 and increased between 1980-81 and 1990-91. But even then, this percentage in 1990-91 was lower compared to 1970-71.

Table 1. Percentage of Land Holdings in Different Categories in India & Punjab

Category of Land Holding	India			Punjab		
	1970-71	1980-81	1990-91	1970-71	1980-81	1990-91
Marginal (<1ha)	50.6	56.8	59.4	37.6	19.2	26.5
Small (1-2 ha)	19.0	18.1	18.8	18.9	19.4	18.2
Semi-Medium (2-4 ha)	15.2	13.7	13.7	20.5	28.0	25.9
Medium (4-10 ha)	11.3	9.1	7.1	18.0	26.2	23.4
Large (> 10 ha)	3.9	2.4	1.6	5.0	7.2	6.0
Total	100.0	100.0	100.0	100.0	100.0	100.0

The marginal and small farms cannot enjoy the benefits of mechanisation through individual ownership. Custom hiring of farm machinery and power is the only means by which they can reap the benefits of farm mechanisation.

Table 2. Number of Land Holding in India and Punjab ('000)

Category of Land Holding	India			Punjab		
	1970-71	1980-81	1990-91	1970-71	1980-81	1990-91
Marginal (<1ha)	35685	50290	63342	517.6	197.3	296.1
Small (1-2 ha)	13409	16007	20048	260.1	199.4	203.9
Semi-Medium (2-4 ha)	10741	12108	13969	281.1	287.4	288.8
Medium (4-10 ha)	7938	8035	7571	247.7	269.1	261.5
Large (> 10 ha)	2763	2142	1706	68.9	73.9	67.2
Total	70536	88582	106636	1375.4	1027.1	1117.5

With the population growing at an alarming rate of 1.8% further fragmentation of the land holdings is continuing. This is clear from Table 2. The number of land holdings in India increased by about 51% between 1970-71 and 1990-91. In case of Punjab where agricultural mechanisation has made maximum progress, the total number of holdings decreased between 1970-71 and 1980-81 and then it increased during the next decade. This number, however, remained lower in 1990-91 compared to 1970-71. Mechanisation seems to prevent or slow down fragmentation of holdings.

Farming Transformation

Agriculture in Punjab, which is the most mechanised in India, has undergone a remarkable change since Independence. It has been transformed from subsistence farming which was entirely dependant on animals for draft power, to mechanised farming using mainly inanimate power sources like tractors, diesel engines, electric motors, etc. The number of all types of farm machinery grew in leaps and bounds bringing in a phenomenal change in Punjab agriculture. Similar transformation is occurring in Haryana and Western UP. Other states are also showing the trend towards progressive to mechanisation of farming.

Custom Operation of Farm Equipment

Custom hiring of farm machinery was first introduced in Indian agriculture in 1912 when a 30-inch (diameter) steam thresher was used for custom hiring in Layallpur district. These machines were taken to about 10 different places working for 2 or 3 days at each place. The users were charged half rate only i.e. 2 anna a maund (Rs. 3.38 per ton). The output of this thresher was about17 kg/h. Organised move to promote multi farm use of agricultural machinery was made in mid 1960's when Agro-Industries Corporations (AIC) were established in the states. The main role of these AICs was to be a facilitator of farm mechanisation. The major objectives of the AICs were:

— Manufacture and distribution of agricultural machinery and implements as well as of the equipment pertaining to processing, dairying, poultry-farming, fishery and other industries connected with agriculture;

- Enable persons engaged in agricultural and allied pursuits to own the means of modernising their operations or alternatively make available the necessary custom service for this purpose;
- Undertake and assist in efficient distribution of inputs for agriculture;
- Promote and develop industries having a bearing on production, preservation and supply of food; and
- Provide technical guidance to farmers and persons concerned with agroindustries with a view to enabling them to conduct their enterprise efficiently.

The Agro-Industries Corporation in most states setup Agricultural Machinery Service Centers to provide custom hiring and servicing facilities to the farmers as no such facility was available. These centres were run on no profit no loss basis. From 1967 these centers also supplied pump sets, tractors, power threshers and power tiller on hire purchase basis rather than on custom hiring. These centres suffered huge losses on account of difficulties involved in recovery of loans from the farmers. The AICs had to curtail this project and finally these centers were closed.

In the meantime, private entrepreneurs entered the field of custom operation of farm equipment. Owners of tractors and threshers started providing custom service to other fellow farmers for tillage, thrashing, transport, etc. This brought additional income to custom operators and gave access to mechanisation for all groups of farmers irrespective of the size of their holdings. Custom hiring of farm machinery got further boost when Govt. of India, in 1971, launched a scheme to set up agro-service centers all over the Country.

The main aim of this scheme was to provide employment to the unemployed engineers and graduates in Agriculture. The Government of India even paid subsidy on the interest (exceeding 5%) on the capital. As a result of this scheme Agro Service Centers in the northern states of Punjab, Haryana and Rajasthan were established to provide custom hiring services for use of tractor operated machinery; to sell farm equipment and other agricultural inputs; and to provide servicing facilities for agricultural machinery. Many of these centers are functioning even today. One can see their harvester combines moving to Central India in February & March and work their way up to North, moving with the wheat harvest from March to early May.

Rural Credit Services

After independence, major thrust of five year plans was on agriculture and rural development. It was strongly felt that both agriculture and rural sector were short of capital and needed credit support. The farmers and villagers were using moneylenders to meet their credit demand and paying exorbitant rates of interest. So, on 19th July 1969 fourteen leading commercial banks were nationalised with the aim of implementing Government Scheme of rural credit.

These banks introduced an experimental scheme of commercial finance at 4% interest per annum for weaker sections including agriculturists. A Credit Guarantee Corporation was set in April, 1971 to operate a voluntary scheme of granting loans up to 75% of the capital subject to certain limitations. This Corporation was also drawn into agricultural development work through 'lead bank schemes'. These nationalised banks were a stimulant for adoption of farm mechanisation and establishment of Agro Service Centers through their provision of liberal credit which was also made available for the purchase of pumps, tractors, diesel engines, electric motors, combines and agricultural machinery.

These banks helped the farmers to take the benefit of agricultural mechanisation technology. They also contributed to the establishment of network of custom hiring, servicing and sales through Agro Service Centres. Other state organisations like land and seed development corporations, marketing federations, etc. also provided valuable technical and financial help to promote custom hiring of farm machinery and tractors.

Status of Indian Agriculture

Since independence, Indian agriculture has been transformed from subsistence agriculture to modern agriculture. The sources of energy and power, farming practices and technologies and crops have under gone a sea change. This is evident from the data on number of mechanical power units used in Indian agriculture. Number of tractors, diesel engines, electric motors, power tillers and self-propelled combines has increased by manifolds during the second half of the twentieth century.

The change in number of farms in different category does not show a uniform trend. This also applies to the average size of holdings in different categories. However, the number of tractors in each category increased over the 12 year period and area per tractor reduced. The marginal and small farmers and 'not tractor owning' farmers in semi-medium and medium categories apparently got 'tractor service' from the tractor owning semi-medium and medium farmers as the area/tractor in these two categories was the smallest and spare 'tractor capacity' was available. In a situation where the use of draught animals has almost disappeared, the vast number of farmers in marginal, small, semi-medium and medium categories who not own tractors continue to operate their holdings with hired power and machinery.

Influence of Infrastructure

The infrastructure needed for the success of custom hiring operations includes all weather roads, servicing network and spatial distribution of machinery ownership. In many parts of northern region adequate infrastructure including repair and maintenance facilities is available and this has lent a great support to the spread and success of custom hiring of farm machines. Agro Service Centers, Co-operative Societies and other public sector undertakings like agroindustries corporations, markfeds, etc. have further augmented the availability of custom hiring services in the region.

Custom Hiring Rates

The custom hiring rates prevailing in Punjab in 2002-2003 are given in Table 3.

Table 3. Custom hiring rates in Punjab

Sr. No.	Operation	Rate (Rs)
1.	Cultivator x 2 and bund making	850 /ha
2.	Harrow x 2	1200 /ha
3.	Planker x 1	288 /ha
4.	Seed drilling	672 /ha
5.	Puddling (Cult x 2 + Planker x 2)	2,280 /ha
6.	Potato planting	1,200 /ha
7.	Potato digging	480 /ha
8.	Straw combine (@ 450/trailer)	2,160 /ha
9.	Reaper	1,920 /ha
10.	Threshing	2,400 /ha
11.	Fertilizers broadcasting	360 /ha
12.	Combine Harvesting of Paddy or wheat	960 /ha
13.	Leveling	600 /ha
14.	Stubble shaver	300 /ha
15.	Spraying	7 /tank
16.	Transportation up to 5 km	300/round trip
17.	Transportation for 5 km to 20 km	500/round trip

Social and Economic Impact

In northern region, more than 80% of the operational land holdings are either marginal (<1 ha), small (1-2 ha) or semi-medium (2-4 ha) in size. Investment capacity of majority of the farmers in these categories of land holdings is poor.

These farmers can not own expensive farm power units and machinery. However, they are making use of modern technology like combine harvester, tillage equipment and planting/sowing machinery through custom hiring. This has helped them improve the timeliness of operation, to increase land productivity and increase economic returns. As a result the living standard of majority of villagers in the northern region has improved.

The size of tractors owned by the farmers of northern region, especially Punjab, varies over a wide range from 20 to 60 hp. Though majority of the tractors are in the range of 35-40 hp (51.8%) yet more than 9% tractors are of more than 50 hp. These large size tractors are mainly used for custom hiring work relating to combine harvesting.

Diversification of Agriculture and Custom Hiring

Diversification is the new mantra for sustainable agricultural development especially in paddy-wheat belt. Diversification means diverting a large area under wheat and paddy to other crops.

Machinery needed for sowing, planting, transplanting, plant protection, harvesting and product recovery is highly crop specific. Thus, diversification would require use of a vast variety of additional equipment for these operations on limited area in the initial stages, making it highly uneconomical on ownership basis.

However, custom hiring through private entrepreneurs or co-operatives will help to increase annual use of these equipment thereby making them economical. Thus, custom hiring of specialised farm equipment for replacement crops can greatly facilitate diversification of production agriculture.

Diversification of agriculture, need for and introduction of new machines and the trend among the farmers to use increasingly larger tractors will vastly expand the scope for custom hiring of farm equipment because in future multi farm use will be the only way to keep the operating cost of farm equipment at a reasonable level. Listed below are examples of farm equipment which have done well in the field in recent years and which are likely to be used on large scale in the future years.

— No till drill
— Pulverising roller cum puddler
— Rotavator
— Strip till drill
— Ridger planter
— Multi crop drill
— Sugarcane planter
— Sugarcane stripper
— Sunflower thresher
— High clearance sprayer
— Straw combine

Most of these machines will prove uneconomical for single farm operation. Under Indian conditions, a majority of the farmers can benefit from the technological superiority of the listed equipment and more such developments in future only through custom hiring.

References

Balasankari, P.K. and Salokhe, V.M., *A case study of tractor utilization by farmers, Coimbatore district,* India. Agricultural Mechanization in Asia, Africa and Latin America 30(3), 14-18, 1999.

Brar, S.S., *Dynamics of agricultural production pattern under varying input output price in Punjab,* Unpublished Ph.D. (Agril. Economics) thesis, PAU, Ludhiana, 1998.

Dhillon, B.S. and Sidhu, D.S. "Economics of custom hiring of agricultural machinery in Punjab – a few case studies", Agricultural Engineering Today, 11(2) 15-19, 1987.

Grewal, S.S., *Impact of mechanization (Tractorization) on labour employment and labour productivity in Punjab,* Unpublished Ph.D. (Agril. Economics) thesis, PAU, Ludhiana, 1972.

Hossain, M.A., *Estimation of demand for major agricultural inputs in Punjab agriculture*, Unpublished Ph.D. (Agricultural Economics) thesis, PAU, Ludhiana, 1988.

Joshi, H.C., *Pattern of tractor power utilization in a fodder farm: A case study*, Agricultural Mechanization in Asia, Africa and Latin America, 29(3). 39-41, 1998.

Kalra, V., *Capital Investment and credit transaction in various farm size in Punjab*, Unpublished M.Sc. (Agril. Economics) Thesis. PAU, Ludhiana, 1996.

Mathur, R.N., *The Andhra Pradesh State Agro-Industries Corporation Ltd.*, Agricultural Engineering Today, 1(7-8), 11-14, 1977.

Morris, J., "Estimation of tractor repair and maintenance costs", *J. Agricultural Engineering Research* 41, 191-200, 1988.

Pandey, M.M. *Energy efficient implements package & needs for their custom hiring* - Agricultural Engineering Today, 24(6), 1-21, 2000.

8

Long-Term Mechanisation Strategy

Mechanisation has been well received the world over as one of the important elements of modernisation of agriculture. It is now recognised that availability of mechanical power and improved equipment has enabled states like Punjab and Haryana to achieve high levels of land productivity.

The results of the survey conducted under the project "Study Relating to Formulating Long-Term Mechanisation Strategy for each Agro Climatic Zone/State" confirm that in those states where agricultural mechanisation has made good progress, its benefits are being shared by all farmers irrespective of the size of their operational holdings and whether they own tractors and machinery or not. However, the progress of mechanisation in most of the states has been slow and its benefits of timely and precise operations, efficient use of costly inputs like seed, fertiliser, plant protection chemicals, limited water resource, etc. are not reaching the majority of farmers in full measure.

During the course of project implementation this matter received deep consideration. It resulted in the identification of some issues which need to be attended to if a more even spread of mechanisation and the policy goal of modernising Indian agriculture have to be achieved. These issues have been given due consideration in making the recommendations on the strategy for mechanisation of agriculture.

Annual Investment on Agricultural Machinery

The annual investment on agricultural machinery including tractors, power tillers, engines, etc., is more than Rs 50,000 crores compared to the combined annual investment of Rs 24,000 crores on fertilisers, certified seeds and plant protection chemicals. Due to poor information and inadequate guidance, the farmer often makes wrong selection of machinery leading to improper investment. The poor management of power and machinery create financial problems for the farmers. This requires:

- — institutional arrangement for giving advise and guiding the farmer in the selection of farm power units and machinery;
- — extensive arrangements for training to young farmers in farm machinery management including its operation and maintenance;
- — active information service to keep the farmer informed of the developments in mechanisation including availability of new and better farm equipment for different applications.

Institutional Frame Work

Institutional arrangements and programmes at the National and State levels are grossly inadequate to plan and give proper direction to the development of agricultural mechanisation. To overcome this weakness, the following is recommended:

- — There should be an Agricultural Mechanisation Board at the Center under the Chairmanship of the Minister of Agriculture. Its function should be to take policy decisions at the national level, review and revise mechanisation strategies, monitor the programmes of mechanisation and make such interventions as may be required to keep the programmes in line with general agricultural goals and policies. Other concerned departments and organisations of the central government including Indian Council of Agricultural

Research, few state governments (by rotation), industries, research organisations, State Agricultural Universities (SAUs), progressive farmers and experts on agricultural mechanisation should be representedin the Board.

- There should be Agricultural Mechanisation Councils in the states which should have the same functions as the Agricultural Mechanisation Board in respect of the states and which should be chaired by the Ministers of Agriculture. The composition of the State Councils should be on the same pattern as of the National Board involving other departments and organisations of the state governments with adequate representation of the industries, progressive farmers and Agricultural Universities.
- The programmes of Agricultural Mechanisation at the Center should be headed by a Commissioner of Agricultural Engineering & Technology who should be directly responsible for the planning and monitoring of agricultural mechanisation programmes at the national level and who should oversee testing and training activities of the FMTTIs. The Commissioner will be responsible for collection and analysis of agricultural mechanisation data and to report the findings to the Agricultural Mechanisation Board. He will represent the Board in all State Councils. The Agricultural Mechanisation Commissioner should be assisted by two Joint Commissioners (Agricultural Machinery) and other professional and administrative staff.
- Director of Agricultural Engineering and Technology should head the agricultural mechanisation programmes at the state level. He should be a qualified professional. In small states, Additional Director of Agricultural Engineering & Technology may head this programme. The responsibilities of the Director Agricultural Engineering at the state level should be the same as that of the Commissioner (Agricultural Engineering and Technology) at the Center. He should

be responsible for coordinating agricultural engineering programmes including mechanisation with the help of zonal and district level staff and the agricultural engineers in Krishi Vigyan Kendras (KVKs). He should organise regular interaction between progressive farmers, SAU experts, representatives of banks and industries and facilitate group visits by farmers and manufactures of farm machinery to national and international exhibitions and other events in which updated information on agricultural machinery is available.

Testing and Training of Agricultural Engineering

— Selected SAUs and Research Institutes with strong agricultural engineering component should be supported by DOAC to develop facilities and undertake farm machinery testing for the purpose of Bureau of Indian Standards (BIS) certification, to help the manufactures to improve quality of equipment and to provide information on machinery performance to DOAC and other organisations. The tractor testing should remain exclusively with FMTTIs Centers. This will greatly increase the DOAC capability for expeditious evaluation of farm machinery and help the industry too.

— The State Directors of Agricultural Engineering, SAUs, State Agro-Industries Corporations, FMTTIs, ICAR Institutes having Tractor Training Centres (TTCs), KVKs and tractor and farm machinery dealers should be made responsible for training tractor and machinery operators. This will require that a qualified agricultural engineer is posted in each KVK. The facilities at selected SAU's and ICAR Institutes should be improved so that these institutions, together with FMTTIs, can offer high quality training to the mechanics, farm machinery fabricators and extension workers in agricultural engineering. The Extension Education Institutes of

DOAC and MANAGE, Hyderabad should have staff and facilities for training middle and senior level extension staff in agricultural engineering.

— Orientation/refresher courses for the functionaries of the above listed organisations should be organised at regular interval with the cooperation of Agricultural Universities and ICAR Institutes so that the trainers remain informed of and in touch with the improvements in farm mechanisation hardware and efficient farm mechanisation management practices.

Research and Development

— Agricultural mechanisation research and development programmes in the SAUs and ICAR Institutes should be strengthened. Agricultural Universities and ICAR Institutes having sufficient manpower and facilities for research and development in the field of mechanisation should be identified and assigned the responsibility of research and development for different agro-economic zones. Other Agricultural Universities, ICAR Institutes, KVKs and NGOs engaged in agricultural mechanisation programmes should be actively associated with the evaluation and demonstration of the output of research and development.

— The programmes of Front Line Demonstration (FLD) of farm machinery should be strengthened.

— Research and Development (R&D) programmes should be reoriented to deliver outputs which are needed by the farmers and are marketable. R&D funding should be generous and adequate to produce successfully tested and well engineered equipment.

Operational Holdings

Due to continuing fragmentation, the average size of operational holdings is shrinking in all States and the percentage of marginal, small and semi-medium operational holdings is increasing. This is making individual ownership of agricultural machinery progressively more difficult. It is recommended to:

— encourage custom hiring operation of tractors, power tillers and farm machinery through training, financial incentives, subsidised loans and adequate financing to allow procurement of high capacity equipment by the custom operator to ensure sufficient turn over and income;

— contract farming should also be encouraged in which case the farmer will grow specific crops suiting to the requirement of the persons/ organisations entering into contract farming. He would also suggest the full package of practices including improved implements and technology for production of specific crops.

— organise farm machinery cooperatives at selected locations and utilise the experience for establishing such cooperatives in all states in sufficient numbers.

Limited Range of Equipment

The R&D programmes in India have so far served mainly the rice-wheat cropping system. The range of the equipment is not wide enough to facilitate diversification of agriculture. Equipment including tractors and prime movers for mechanisation of hill agriculture and production of fruits and vegetables is not commonly available. The R&D programmes in future should focus on the development of farm machinery and power units for:

— precision and protected agriculture,

— hill agriculture,

— horticulture, cash and plantation crop,

— recovery and management of crop residues,

— non-farm applications like efficient rural transport, maintenance of village roads, etc.,

— sample equipment like hill side tractors, tool carriers, precision planters, transplanters, harvesters and pickers for sugarcane, cotton, fruits, etc., high capacity mowers, swathers and balers and mechanical handling equipment should be imported and made available to R&D Institutes for proper evaluation and integration with our requirements.

Energy Conservation

Land and water, the two natural resources, are limited and need to be managed in such a way that their quality and productivity are maintained and possibly improved. Energy conservation has become important because the diesel prices are rising and the supply of grid power to villages remains limited. It is recommended that:

— The farmers should be encouraged to incorporate crop residues into the soil to improve its organic matter content and both animal drawn and tractor operated equipment should be developed and made available for this purpose.

— Periodic sub-soiling should be practised to avoid hard pan formation and custom operators should be encouraged to add this operation to the services offered by them.

— Equipment for forming raised beds, precision levelling and for other such operations which help to improve water use efficiency should be made available through custom operators or farm machinery cooperatives.

— Energy saving practices like zero till planting, strip till planting, use of machine combines and throttle down - high gear operation of tractors should be demonstrated and promoted.

— Micro-irrigation system viz. sprinkler and drip irrigation system should be encouraged to save energy and improve water utilisation efficiency.

Agricultural Tools and Machinery

Fabrication of agricultural tools and machinery is often done by semi-skilled workers and without proper tooling. This is more so in the case of hand tools and animal drawn equipment. The poor quality of agricultural equipment puts the user-farmer and custom operator at a permanent disadvantage. There is vast opportunity for exporting animal drawn equipment and power equipment in small and medium range to the countries in Asia and Africa. The poor quality of agricultural equipment has prevented the Indian manufacturers from using this opportunity. It is recommended that:

- selected SAUs, Agricultural Engineering Institutes of ICAR and ITIs should be supported to develop competence and offer intensive training to young technicians and artisans in the fabrication and manufacture of farm tools and machinery;
- farm machinery manufacturing is reserved for small scale industries. Organised industries should be allowed to manufacture farm machinery;
- while the tractor testing and training centres should continue to test tractors and power tillers, zonal farm machinery testing centres should be established in SAUs and other selected organisations and only successfully tested equipment should be recommended to the users.
- farm machinery manufacturers training programmes and manufacturers business meets may be organised to create awareness among the manufacturers about the production of quality agricultural equipment and machinery. Facilities may be extended to them for testing and evaluation of their equipment from the product development point of view. Incentives and awards should be given to the manufacturers for producing quality implements/machinery.
- FMTTIs/SAUs/ICAR Institutes and other R&D organisations should offer technical support and testing facilities to the farm machinery and processing equipment manufacturers to improve the quality of their products.

New Developments and Improvements

In the absence of adequate interaction with research organisations and industries in technologically advanced countries, R&D groups and industries in India are not able to incorporate new developments and improvements in their designs and products. It is recommended that:

— industries and R&D organisations should be allowed duty free import in limited numbers of farm machinery from other countries for the purpose of adaptive research and product improvement;

— Government should facilitate group visits by machinery manufacturers and research workers to international farm machinery and tractors shows and exhibitions.

Low Level of Mechanisation

There are many regions in which the farmers continue to use country plough and inefficient hand tools because little effort has been made to introduce improved animal drawn implem-ents and hand tools in these areas. It is recommended that:

— special schemes should be launched to popularise improved animal drawn implements and hand tools in areas which have backward agriculture;

— financial incentive should be provided for replacement of inefficient equipment by improved tools and implements till this transition picks up speed.

— to accelerate the growth of agricultural mechanisation in these regions, special incentives and high rate of subsidy on different types of agricultural machinery should be provided.

— each State Government should themselves draw their annual plan for promoting agricultural mechanisation in their regions.

Bank Financing

Almost all tractors are purchased through bank financing by mortgaging the land. This arrangement discriminates against the marginal and small farmers and unemployed youth who are likely to be most keen to take up custom operation. It is recommended that:

- bank loans for purchase of tractors and agricultural machinery should be given on the basis of feasibility study to ensure financial soundness of the proposal and the purchased equipment should be hypothecated to the bank;
- purchase of second hand tractors should be financed on the same basis, viz., financial soundness.

Animal Draught Power

Draught animals remain an important source of power for a large number of farmers, particularly in regions where the land is submerged during rains, the soils create problem of traction or the terrain is hilly. It is recommended that:

- few R&D institutions should be identified to concentrate on the development of better animal drawn implements and machines for different crops and operations and adequate resources should be provided for this purpose.

Power and Fuels Supply

The grid power supply to the rural sector is limited, uncertain and of poor quality. This erodes the farmer's capability to regulate the timing of farm operations and leads to mismanagement of power and resources like water. Rising fuel prices are creating economic problems for the progressive farmer. It is recommended that:

- surplus biomass should be utilised as energy resource to supply fuel and power to production agriculture and for the development of agro-industries in the rural sector;

— an efficient package of equipment and practices should be developed to allow proper management of biomass which is now destroyed through uncontrolled burning, for use as energy material;

— priority should be given to the development of equipment for collection, densification, handling, transport and storage of surplus crop residues for use as energy resource;

— priority should be given to the development of efficient and economically viable conversion systems to produce fuel gas and liquid fuel from surplus crop residues and to decentralised power generation in production catchments where surplus crop residues are available.

Maintenance of Agricultural Equipment

In the initial stages of introduction of new machines, inadequate repair and maintenance facilities dampen the enthusiasm of the user farmers. To ensure smooth spread of agricultural mechanisation to all regions, training for young technicians and artisans in the repair and maintenance of agricultural equipment should be provided in KVKs, ITIs and by Non government Organisations (NGOs) with financial support from the State Government.

Generally, the agricultural mechanisation data is based on the Livestock Surveys conducted in every State and these data are unreliable and unrealistic due to possible human errors in enumeration of technical details of machinery by nontechnical personnel. It is suggested that for collection of reliable data, enumeration of mechanisation data may be done by technical personnel. District-wise Computerised Data Banks may be set up to record all relevant data on production and sale of different types of agricultural machinery. Such information should be published annually for updating of the data for formulating future strategies as well for analysing the impact of the mechanisation programmes.

References

Patel, A.R., *Agro Service Center Scheme: Need for Monitoring*, Agricultural Engineering Today, 4(4), 5-7, 1980.

Peter, E.C. *A proposal to accelerate farm mechanization*, AET, 17(3-4), 68-72, 1993.

Randhawa, M.S., *A history of agriculture in India*, All Volumes, Indian Council of Agricultural Research, New Delhi, 1980.

Sandhu, P.S., *A study of change in structure, resourceuse efficiency and factor share in Punjab agriculture*, Unpublished Ph.D. (Agricultural Economics) thesis, PAU, Ludhiana, 1985.

Sethi, K. *Land use pattern and demand for tractors*, Unpublished M.Sc. (Agricultural Economics) Thesis. PAU, Ludhiana, 1993.

Sharma, M., *A quantitative analysis of demand for tractors in India with particular reference to Punjab*, Unpublished M.Sc. (Agricultural Economics) Thesis. PAU, Ludhiana, 1998.

9

Safe Operation of Farm Machinery and Equipment

Tractors, threshers and chaff cutters are the main cause of accidental deaths on farms. Over the years, many farmers, farm workers and others living on or visiting farms, have been killed or seriously injured falling from moving tractors, being run over by tractors, or being crushed when a tractor rolls sideways or backwards.

Spot the Hazard

Regularly check for hazards relating to tractors, attached implements and field conditions. Hazard areas could include mechanical parts, operator training, other people, work procedures, unsafe jacking, climatic conditions, chemicals used, uneven terrain, and any other potential causes of an injury or a hazardous incident. Keep a record to ensure identified hazards are assessed and controlled.

Risk Assessment

Once a potential hazard has been identified, assess the likelihood of an injury or hazardous incident occurring. For example, risk to children playing near a tractor will vary, depending on what the tractor operator is doing, how close they are to the tractor and whether the operator knows they are there. Consider ways of minimizing risk.

Make the Changes

- Read and follow safety procedures in the manufacturer's manual.
- Ensure an approved cab or rollover protective structure (ROPS) is fitted.
- Fit and use a seat-belt on tractors with ROPS.
- If there is a risk from falling objects, fit a fall-on protective structure (FOPS).
- To reduce risk of back strain, fit a seat with side restraints and a backrest.
- Wear hearing protection, and remember, not all tractor cabs are sound proof.
- Keep children away from tractors and machinery.
- Remove starter keys when tractors are not in use.
- Have an up-to-date maintenance schedule.
- Follow safe maintenance and jacking procedures.
- Ensure that operator is properly trained for each type of tractor work.
- Always mount and dismount on a tractor's left side - to avoid controls.
- Adjust the seat so that all controls are safely and comfortably reached.
- Keep all guards in place, including the power take-off (PTO).
- Operate the self-starter from the operator position only.
- Never carry passengers.

Operating a Tractor

- Drive at speeds slow enough to retain control over unexpected events.
- Reduce speed before turning or applying brakes.
- Watch out for ditches, logs, rocks, depressions and embankments.

— On steep slopes, without a trailed implement, reverse up for greater safety.
— Engage the clutch gently at all times, especially when going uphill or towing.
— Use as wide a wheel track as possible on hillsides and sloping ground.
— Descend slopes cautiously in low gear, using the motor as a brake.
— Never mount or dismount from a moving tractor.
— Ensure the park brake is on and operating effectively before dismounting.
— Take short breaks regularly when working long hours.

Towing Implements

— Fit attachments according to the manufacturer's instructions.
— Always attach implements to the draw bar or the mounting points provided by the manufacturer.
— Never alter, modify or raise the height of the draw bar unless provided for by the manufacturer.
— Regularly check safety pins on towed lift-wing implements, to ensure they are not worn.
— Ensure all guards on towed implements are in place before operating.
— Never hitch above the centerline of the rear axle, around the axle housing or to the top link pin.
— Never adjust or work on implements while they are in motion.
— Never attach implements unless the PTO shaft is guarded.
— When parking, always lower the three-point linkage and towed implement.

Maintenance of Tractor

People have been killed and seriously injured doing maintenance and repairs to farm tractors. Major hazards can occur when tractors are jacked and wheels are removed without safe working procedures. These risks are magnified on soil. Regular workshop maintenance of farm tractors and trailed implements can prevent hazardous incidents in the field.

Spot the Hazard

When planning tractor maintenance, check the right equipment is available for safe jacking, removal of wheels and other tasks. People doing the job should be experienced, and there should be agreed safe procedures. Heavy lifting and carrying can cause strain injuries. Children should be kept away from tractor workshops. Field repairs present specific injury risks.

Assess the Risk

The greater the risk of an injury or a dangerous incident occurring, the more urgent the need for changes to be made to minimize or eliminate the risk.

Make the Changes

Here are some ways of improving tractor maintenance safety.

- Routinely adjust brakes, clutches and drives, according to the manual.
- Ensure steering, exhaust system and brakes are in top condition.
- Stop the motor before re-fuelling, servicing or greasing and, if possible, wait until the engine is cold before re-fuelling.
- Never remove or replace belts while pulleys are under power.
- Keep steps and working platforms free of grease and oil to avoid slips and falls.
- If the engine overheats, allow time for it to cool off before removing the radiator cap.

Jack Tractor

- Jack on a flat surface, ideally a concrete floor.
- Avoid working alone. Ask somebody with training and experience to help.
- Refer to the manufacturer's manual on safe jacking, or seek professional advice.
- Where jacking points are not identifiable, jack from the lowest possible point.
- Use jacks that comply with applicable standards.
- Use vehicle stands that comply with applicable standards, and are designed for the load to be suspended.

Blocks and Chocks

- Ensure wooden blocks for jacking are of hardwood, e.g. jarrah or karri, with a surface area that will support the tractor's weight on soft soils.
- Chock all wheels that will remain on the ground, using big wooden chocks at the front and rear of each wheel. Don't use rocks; they're too unstable.
- Chock all wheels on articulated vehicles to stop them twisting sideways during jacking.
- Before jacking, apply brakes, place in gear - or automatic park - and switch ignition off.
- Stay clear of the tractor while operating the jack.

Removing Wheels

- Loosen wheel nuts before the wheel is off the ground, to avoid any movement that could dislodge the tractor.
- Before removing a tractor tire from a rim, release all water and air pressure.
- To avoid serious injuries, it is recommended that work performed on split rims be done by the professional. Therefore, farm workers should not work on split rims.

— Never jack more than one wheel off the ground at a time in the field.

— If both rear wheels have to be removed, work on a flat, level concrete floor, in the workshop.

— When removing rear wheels, ensure the front wheels are immobilised by fixing wedges between axle and body.

Grain Harvesting Equipment Safely

Good safety habits are vital for anyone who operates a combine, corn picker or other grain-harvesting machine. Failure to observe safety practices can be fatal! However, constant alertness is also necessary to prevent machinery accidents-accidents that often happen in spite of machinery that is designed for safety. Machinery operators are not in top physical or emotional condition when they are tired, ill, worried, angry, or have their minds on something else. Accidents are most likely to happen under these conditions.

The combine operator is responsible not only for his safety but also for the safety of others who may be working on or just be near the machine. The operator must be aware of hazards and remain alert to situations that are potentially dangerous. This includes pre-operational checks, starting, transporting, towing, operating, field repair and maintenance and stopping the combine.

Hand Signals

A set of hand signals has been endorsed by several safety institutions. Since spoken instructions are very difficult to hear over the sounds of a combine, knowledge of hand signals can be extremely helpful to the operator when maneuvering a combine, especially in tight places.

Safety Before Starting

— Before attempting to operate a combine, study the operator's manual. It has information on general safety rules, plus specific safety recommendations for the particular machine. The more you know about the combine, the better prepared you will be to safely operate it.

— The exhaust fumes from a gasoline or diesel engine are very poisonous. If the combine is run inside a building, be sure to open the doors to provide good ventilation.
— Always clean the combine before starting. Trash around the exhaust system can cause fires. Oil, grease or mud on ladders or the platform can cause serious falls. If the combine is equipped with a cab, clean the glass to provide maximum visibility.
— Check the tire pressure each day. Under-inflation can cause buckling of the sidewall, which can cause dangerous tire failure. Over-inflated tire have a great deal of "bounce" and cause upsets more readily that tires with correct pressure.
— Check the brakes once a week. With hydraulic brakes, make sure that the master cylinder is full of fluid and that no air is present in the lines. Adjust the pedal free travel, if necessary, so that the brakes are engaged with the pedals an equal distance from the floor of the platform. Check the operator's manual for specific instructions.
— Check the threshing cylinder-rocking bar to see it is clear of the cylinder.
— Make sure that all shields and covers are in place and fastened securely.
— Remove or stow all service equipment.
— Always use the handrails and ladders provided on the combine for safe mounting and dismounting.
— Be sure that all PTO covers, safety stands and shields are on the machine before taking it to the field.

Starting the Combine

— Before mounting the combine, make sure that everyone is clear of the machine. Do not allow anyone to ride with you, unless combine is equipped with a passenger seat.

— Before starting the combine:
  — Disengage header drive.
  — Disengage separator drive
  — Place gearshift in neutral
  — Depress clutch pedal
— Be careful when using diesel starting fluid. It is extremely flammable.
— If it is necessary to use jumper cables to start the combine, be careful to avoid sparks around the battery. Hydrogen gas escaping from the battery can explode. Follow the operator's manual instructions for using jumper cables.

Transporting the Combine

— Always keep your mind on the dangers of driving the combine on public roads. Beside maintaining control of the machine, you must watch for obstacles on the road, pedestrians and traffic.
— High speed is the leading cause of accidents. Never drive faster that the road conditions allow for safe operation. Anticipate dangers and slow down to avoid accidents.
— Make sure you are familiar with local traffic laws. Check the safety flashers and small moving vehicle (SMV) emblems to be sure they are clean and visible.
— Always lock the brake pedals together. If the combine is not equipped with locking mechanism, be sure to depress both pedals at the same time evenly. Applying only one brake, or applying one harder than the other can cause the combine to swerve and perhaps tip over.
— Be careful when applying brakes when a header is attached to the combine. The added weight up front can cause the combine to tip forward if the brakes are

applied abruptly. Always drive slow enough to allow controlled application of brakes at all times.

- Always check headlights and safety flashers to make sure they are properly adjusted and in working order.
- Put the unloading auger in the transport position. Be certain it is not blocking a safety flasher or SMV emblem.
- On self-propelled combines, never use the header safety support when transporting the machine. Raise the header enough for safe ground clearance, but not high enough to reduce visibility.
- On pull-type combines, always use header support when transporting. Towing at transport speeds can be hazardous because of side forces on the tractor when stopping too quickly. Side forces from slowing a combine too quickly may cause a tractor to skid, especially on loose gravel. Slowing down while turning can cause jack-knifing. Slow down before the corner so the towed combine doesn't get out of control.
- Watch for low power or telephone lines, bridges, buildings and any other obstacles, to make sure you can pass under them safely. Always keep as far to the right of the roadway as possible. Keep a careful watch to see that you have safe clearance on both sides.
- Always sit down when travelling at high speeds or going over rough terrain.
- Be careful when making turns. Make sure that the rear of the combine will clear obstacles when it swings around. Avoid sharp turns. Turning too sharply at high speed can cause the machine to turn over.
- Because the wheels for steering are in the back, self-propelled machines often fishtail when turned too quickly at transport speeds. Steering to the right will whip the rear to the left, and vice versa. Steering suddenly to the right when meeting oncoming traffic

causes the back of the combine to swing out into the path of on coming traffic.

— Slowing or braking too rapidly could cause loss of some steering control (weight on rear wheels). This is most noticeable when driving with a corn head or some other heavy header raised high. In this case, most of the weight will be on the drive wheels. Install rear wheel weights. Keep header as low as possible. Use the variable speed drive or engine throttle to slow the machine. Reduce speed before you need to apply brakes and always lock brake pedals together.

— Never depress the clutch pedal or take the combine out of gear to coast down hill. When the combine is moving it is impossible to shift the transmission back in gear. Always maintain complete control of the combine. The same applies to tractors that are towing pull-type combines.

Towing the Combine

— If the combine must be transported over long distances, it is safer to haul it on a large truck or a special low trailer.

— Never tow the combine at speeds higher than 20 mph.

— Always keep the transmission in neutral or in the "tow" position, if the combine is so equipped.

— Never tow a combine equipped with hydrostatic drive. Towing can cause damage to the drive unit. Instead, haul the combine.

Operating the Combine

— Never operate the combine if you are ill or sleepy. Operating safety depends on alert, efficient handling of the combine.

— Wear safety glasses at all times.

— Wear clothing that fits snugly to avoid catching clothing in moving parts.

— Never let anyone ride on the combine unless it is equipped with a passenger seat. A rider's clothing may become entangled in moving parts, or he may be thrown off the machine.

— Before starting to harvest a field, check it carefully for ditches, fences or other obstacles. Be aware of weather conditions, which present safety hazards.

— Be especially careful when operating on hillsides. Avoid sharp turns that could tip the combine over. Beware of ditches or obstacles-they are doubly dangerous on slopes.

— If grain tank extensions are used, remember that the added weight may make the combine top heavy and more subject to upsets.

— Never travel over 10 mph (16 km/h) with a full grain tank. The added weight makes the combine more difficult to maneuver and easier to upset.

— Always sit down when travelling over rough terrain. A sharp jolt can throw you from the platform or away from he controls.

— Hillside combines are equipped with automatic or manual leveling devices. Hydraulic cylinders act to level these combines on steep slopes. These machines are equipped with a warning signal that indicates when the leveling system has reached its limit. Be especially careful after the device activates.

— When using the steering brakes, always turn the steering wheel before applying the steering brakes. Failure to do so can cause the combine to swerve and turn dangerously.

Field Repair and Maintenance Safety

— Always keep the machine clean. Field trash around the exhaust system can cause fires. Mud, grease or oil on the operator's platform or ladders can cause falls.

— Before lubricating or adjusting the combine, disengage all drives and stop the engine. Never leave the operator's platform with the engine running.
— Make sure that the header drive and separator drive are disengaged before attempting to clean the combine. Never try to unclog the machine with a stick or pole with the machine is running. The stalk rolls on a corn head can pull a 12 foot (3.6 cm) stick through in one second-shorter sticks or stalks even faster-before you can let go.
— On a pull-type combines, always disengage the PTO and turn off the tractor before attempting to unclog, adjust or lubricate the machine.
— Always stop the machine before opening the inspection doors.
— Keep all shields in place. After working on the combine, make sure the shields are fastened securely.
— When operating in very dusty or noisy locations, wear goggles and ear plugs to insure safe visibility and prevent hearing loss. Never wear loose clothing that can become entangled in moving parts.
— Stay clear of moving parts at all times.
— Keep belts and chains properly adjust and aligned.
— Don't rely on the hydraulic system for support when working under the machine header. Always use the stops or supports provided on the machine. If no safety device is provided, block the header securely.
— When adjusting the wheel spacing, make certain the machine is blocked. Never rely on jacks alone for support.
— Always support the reel arm securely when adjustments are being made.
— Be careful when removing heavy parts. Make certain they are held firmly to avoid dropping them. Have someone help you with heavy jobs.
— When operating in dry fields. Install a spark arresting muffler to prevent fire.

— Avoid sparks or open flames when working the battery. Hydrogen gas escaping from the battery may explode.

— When possible always refuel the combine outside the field. Let the engine cool before attempting to refuel and never smoke around fuels.

— Allow the system to cool and remove the radiator cap slowly, turning it until pressure escapes through the overflow pipe. Make sure all pressure is relieved before removing the cap.

— Stay clear of the exhaust system until it cools.

— High-pressure fluid leaks in the hydraulic or diesel fuel system are very dangerous. The leaks can be invisible and still have enough pressure to penetrate the skin. When checking for leaks, use a piece of cardboard. If an injury does occur, seek medical aid immediately.

— Always carry a first aid kit and fire extinguisher on the combine.

Stopping the Combine

To make sure drive units do not cause injury when the machine is started again, do the following when stopping the combine.

— Disengage header drive

— Disengage separator drive

— Place gearshift lever in neutral

— Lower header

— Apply parking brake

— Remove ignition key to prevent tampering or accidental starting.

Baling Hay

Large hay bales, some weighing up to 800 kg, have killed or seriously injured many farm workers. Bales, both round and rectangular, can fall on tractor and forklift operators, topple off stacks and vehicles on workers or bystanders, and collapse when stacks fail.

Spot the Hazard

Look for hazards relating to Handling:

- — Children playing near hay balers, carriers and stackers.
- — Training of operators handling tractors, front-end loaders or forklifts.
- — Use of two-poster tractor ROPS for baling - they offer no operator protection from bales falling back off forks or bale-loading frames. Tractors with cabs, FOPS (fall on protective structure) or four-poster ROPS are safer.
- — Makeshift or poorly fitting bale-loading attachments on tractors and forklifts.
- — Carrying bales too high off the ground.
- — Insufficient counterbalance on tractor or forklift vehicle.
- — Hydraulic control valve should be specific to the front-end loader attachment.

Baler operation:

- — Baler properly connected to the tractor.
- — Adequate safety guards fitted.
- — Nobody allowed to ride on the baler.
- — Prevent others getting too close to the baler.
- — Build-up of loose, combustible material in the baler.
- — Fire extinguisher fitted to the machine.
- — Disengage PTO and apply fly wheel brake prior to making baler adjustments.
- — Stop engine and apply fly wheel brake before repairs or "stringing up" the baler.
- — Extra care and attention when reversing or turning the machine,
- — working at night,
- — loading onto a truck, and unloading.

Stacking:

— Loading and stacking on uneven ground.

— Stacks under or near overhead powerlines

— Stacks of round bales inadequately chocked and border posted.

— Damaged bales at base of stack - e.g., from vehicles, cattle or rodents.

— Unstable heights and loose stacking.

— Bales stacked higher than safe operating height of farm tractor or forklift.

— Children playing on stacked bales, particularly during stacking or unstacking.

— Lack of training, experience and protection for people doing hay baling, stacking and loading.

— Handling more bales than safe for the loader.

Transporting:

— Sturdiness of trailers carrying heavy loads of bales.

— Restraining frames back and front of trailer.

— Hooks fitted so ropes can be used to secure load.

— Roads too close to or below powerlines.

— Rough terrain causing bales to become unstable.

— Safe speeds at all times.

— People riding on loaded hay trailers - highly dangerous.

Assess the Risk

Check each hazard that has been spotted to assess:

— Likelihood - how likely is this hazard to injure someone? and

— Severity - how severe would that injury be?

List all the hay baling and stacking hazards spotted, and number them in order of priority, so that those most likely to cause injury or harm can be tackled first. Because most large hay bales can kill or seriously injure anyone they fall or roll on, any risk of a hazardous incident should be assessed as requiring urgent attention. And as children are the most vulnerable, consider child injury risks top priority.

Make the changes Handling:

— Keep children away from hay baling and stacking operations.
— Make sure operators and handlers are properly trained and physically capable of tasks.
— Consider fall-arrest protection for people working at heights.
— Use tractors with four-poster ROPS, FOPS or cabs for protection against falling bales.
— Replace risky attachments with manufacturer approved attachments.
— Ensure loader or forklift operators transport bales close to the ground.
— Avoid sharp turns and unsafe speeds.
— Make sure vehicle controls are fitted specifically for the attachment in use.

Stacking:

— Stacks should be on firm, level ground, away from fire hazards, sources of ignition, overhead powerlines, dwellings, boundary fences and footpaths.
— Make sure stack and load heights do not exceed the lifting capabilities of the farm handling equipment.
— Big bales should be stacked to a maximum of four bales high.
— High density bales can be stacked up to six layers high.
— Wherever possible, stack big rectangular, square or high density bales by overlapping, to form a stable stack.
— Do not use bale lifting equipment to raise people on or off stacks.

Transporting:

— Do not allow people to ride on stacked trailers.

— Be aware of overhead obstructions, like trees, bridges and powerlines.

— Avoid rough ground that could cause bales to dislodge.

— Ensure loads are adequately secured.

— Do not overload vehicles beyond legal limits.

Tillage Equipment

While the skilled operator of tillage equipment avoids errors with very little conscious thought, accident studies show that hurrying and human error are responsible for or are involved in the vast majority of equipment accidents. An operator must have an understanding of the function, operation and limitations of the equipment he/she is operating and the operator must resist the temptation to be hurried into an accident.

Other Informations

Moldboard plows

— Provide adequate front-end weight for tractor stability in transport and operation, particularly with integral and semi-integral plows. Never pull from any point higher on the tractor than the recommended hitch point.

— Use extreme caution and reduce speed when transporting the plow and the tractor over rough ground.

— Avoid sharp turns at high speeds, especially on slopes.

— On tight turns, avoid swinging rear of plow into fences or other obstacles.

— Turning stops on semi-integral plows limit turning radius. Shorter turns may severely damage plow frame and tractor hitch.

— Never carry passengers on the tractor or permit others to ride on the plow - particularly plows with automatic reset.

— Always lower the plow when not in use or left unattended.

— Lower the plow and securely pin the parking stand before detaching the plow from the tractor.

— Always use proper lighting, reflectors, slow moving vehicle (SMV) emblem, and other safety devices for road travel as required by state and local laws.

— When hitching drawn plows, always use a hitch pin with adequate strength for the tractor-plow combination.

Disk plows

— Integral plows are transported completely raised, and all weight is carried by the tractor 3-point hitch. Adequate tractor front-end weights are required to offset the plow weight.

— When the transporting on a road or highway, always display SMV emblem and use lights and reflectors as required by state and local regulations.

— Semi-integral plows are quite long and caution must be used when turning to prevent swinging the plow into fences or irrigation ditches.

— Reduce speed when transporting over rough ground, and avoid quick, sharp turns at high speeds.

— When transporting semi-integral or drawn plows, always install cylinder locks to prevent accidental lowering of the plow. Relieve the load on hydraulic cylinders before starting to transport.

— Lower the plow to the ground or install hydraulic cylinder locks when the plow is not in use.

— Watch for other people when raising, lowering, or indexing the plow.

— Never permit anyone to ride on the plow, and allow only the driver on the tractor. Do not permit children to play on or near the plow either when parked or in operation.

— Lower the parking stand and securely pin it in place before detaching integral or semi-integral plows from the tractor.

Disk illers

— Keep SMV emblem clean and prominently displayed. Do the same with reflectors and warning lights as required by state and local regulations.

— Never allow anyone but the operator to ride on the tractor.

— Never ride or permit others to ride on the tiller.

— Lower the tiller to the ground when not in use.

— Secure the machine in the raised position by installing safety locks or hold-up pins when servicing or cleaning it.

— Disk blades are extremely sharp; be very careful when working or making adjustments in the disk area.

— Never walk close beside the rear wheel when the tiller is in operation. A sudden imbalance of forces could cause this wheel suddenly to jump to the left.

— Never grease, oil, or adjust the tiller while it is in operation.

— Escaping hydraulic oil under pressure can cause serious personal injury and infection. Therefore, be sure all connections are tight and that oil lines are undamaged. Always relieve hydraulic pressure in lines before disconnecting hoses.

Chisel plow

— Reduce speed when transporting chisel plows over rough or uneven terrain.

— Use lock-up straps or transport locks when transporting a chisel plow.

— Be sure wings are locked in the folded position before travelling.

— Use proper lights, reflectors, and a clean SMV emblem when transporting equipment on road or highway.

— The transport width of most folding-section plows exceeds maximum width of normal vehicles. Therefore, use extreme caution when meeting other traffic, to avoid collisions and the possibility of transport wheels dropping into holes, drains, or ditches along the road edge.

— Allow only the operator to ride on the tractor.

— Never allow passengers to ride on the chisel plow.

— Never allow anyone to stand or work near the chisel plow when it is in operation, particularly when raising or lowering outriggers.

— Do not permit children to play on or near the chisel plow during operation or storage.

— Provide adequate tractor front-end ballast for stability in transport and operation, especially with integral models.

— Be particularly careful of escaping hydraulic fluid, which can penetrate the skin and cause serious infection or reaction if not given immediate medical treatment.

— Do not unhitch from the tractor or store a chisel plow when outriggers are in the raised position.

— Make sure raised outriggers will safely pass under power and telephone lines.

Disk harrows and offset disks

— Always lock safety lock during transport, if the disk is to be left raised for an extended period, or while working on the machine.

— Never depend on tractor hydraulic pressure to carry harrow weight in transport-use safety lock, and relieve pressure in cylinders.

— Lower integral harrows to ground each time tractor engine is shut off, and any time harrow is being serviced or repaired. If it must be raised for repairs, securely block the frame to prevent accidental lowering.

— Always use lights, reflectors, and SMV emblem when transporting, day or night.

— Lock the tractor drawbar in fixed position when transporting wheeled disks.

— Never transport a disk harrow on its own wheels at more than normal tractor speed, and considerably less than that on rough or uneven ground.

— Never clean, adjust, or lubricate the harrow while it is in motion.

— Wear protective gloves when working with or near disk blades.

— Hydraulic fluid escaping under pressure can penetrate the skin and cause serious infection or reactions. Never use hands to locate the source of a small leak which may be nearly invisible. Obtain immediate medical attention if injured by escaping hydraulic fluid.

— Park or block the harrow so it cannot roll when unhitched.

— Make sure wings are securely locked in transport position before moving the harrow.

— Large disk harrows exceed normal vehicle width, so be particularly careful to avoid collisions when meeting other vehicles on the road. Avoid dropping wheels of tractor or harrow into holes, drains, or ditches along the road.

— Provide adequate tractor ballast for front-end stability and to prevent excessive slippage.

— Never allow anyone to ride on tractor drawbar or harrow in operation or transport.

— Never allow anyone but the operator to ride on the tractor.

— Lower the machine or install safety lock when storing a disk harrow.

— Never permit children to play on or near a disk harrow while it is in operation, transport, or storage.

— Stand clear of harrow wings during folding or unfolding.

— Remove spring-loaded scrapers in proper order to avoid personal injury. Use care in relieving any spring under tension or compression.

— Do not make sharp turns with blades down.

Field cultivators

— Never exceed recommended transport speed for the cultivator used. If speed is not stated, do not exceed maximum tractor speed.

— Reduce speed for turning and travel over rough or uneven ground.

— Use transport locks and relieve pressure in cylinders when transporting field cultivators. Do not depend on hydraulic pressure to carry the weight. Always lock wings in transport position and relieve pressure in cylinders.

— Never walk or work under wings when they are in the folded position.

— Follow state and local regulations regarding lights, reflectors, SMV emblem, and maximum width when transporting on roads or highways.

— Transport width of most field cultivators exceeds normal vehicle width. Therefore, use extreme caution when meeting other vehicles and avoid the possibility of dropping tractor or- implement wheels into holes, drains, or ditches along the road edge.

— Never permit anyone to ride on the tractor drawbar or cultivator in transport or operation, or to stand near the machine while it is operating- particularly when raising or lowering wings.

Toothed harrows

— Provide sufficient front-end weight for safe, stable operation and transport.

— Use widest practical wheel tread to improve tractor stability, especially when working on steep slopes.

— Reduce implement to narrowest possible width for transport on roads or highways.

— Lock wings or folded sections securely in place before transporting equipment.

— Use lights, reflectors, and SMV emblem as required by law when transporting equipment-day or night.

— Schedule moves for least hazardous periods; avoid transporting equipment on busy roads, during peak traffic periods, or after dark.

— Never transport wheeled harrows, harrow carts, or field conditioners at more than tractor transport speeds; transport considerably slower on rough or uneven terrain.

— Never make sharp turns at high speeds.

— Never allow anyone to ride on the tractor but the operator.

— Never allow anyone to ride on the tractor drawbar or implement in operation or transport.

— Always stop implement and tractor engine to adjust, repair, or lubricate.

— Lower implements to the ground before stopping tractor engine, before servicing or repairing equipment, or at any time the machines are left unattended.

— Never make extremely short turns with drag-type eveners which could foul tractor tires.

— Never park implements where they could be hidden by tall or growing crops, grass, or weeds.
— Keep PTO shaft properly shielded. Never get off tractor without disengaging PTO and stopping engine.

Packers

— Always use reflectors, lights, and SMV emblem as required when transporting equipment-day or night.
— Provide adequate front-end weight for tractor stability in operation and transport of integral harrows. Use maximum allowable front ballast if operating integral harrows in lower gears.
— Never exceed normal tractor speed when transporting drawn roller harrows, and drive considerably slower than that on rough or uneven ground.
— Do not transport roller packers over hard-surfaced roads-use carriers.
— Never permit anyone but the driver to ride on the tractor.
— Never permit anyone to ride on the tractor drawbar or implement during operation or transport.
— Install transport lock pin before storing, transporting, or parking drawn implement; do not depend on hydraulic pressure to support the weight. Lower machines to the ground whenever the tractor engine is shut off.
— Pin tractor drawbar in center before transporting.
— Never lubricate, adjust, or repair the implement while it is in motion or the tractor engine is running.
— Always raise spring teeth before lowering roller harrow to the ground for parking or storage.
— Park or block the implement to prevent rolling when it is disconnected from the tractor.
— Never try to lift or support the roller harrow on the spring teeth for service or repairs.

Undercutters

— Use the SMV emblem, lights, and reflectors as required by law for transporting equipment on roads or highways.

— Install the transport safety lock and relieve hydraulic pressure in the cylinders when transporting.

— Be certain wings are securely locked in the folded position before transporting.

— Limit transport speed as recommended--15 miles an hour (24 Km/h) for some machines, even less on rough or uneven terrain.

— Never ride or allow others to ride on the machine during operation or transport. Allow only the driver on the tractor.

— Do not unhitch from the tractor or store the machine with wings in the folded transport position.

— Sweeps are sharp-watch out for them when wings are folded for transport or when servicing, adjusting, or repairing the plow. Always position wrenches to pull away from sharp edges or corners.

— Never stand with feet under blades while making adjustments or during maintenance. Be extremely careful while working within the implement frame.

— Do not stand or walk on the plow frame, or under wings when they are folded for transport.

— Stand with both feet on the same side of the tongue when hitching or unhitching.

— Do not allow children to play on or near the plow.

All-Terrain Vehicles

The term 'Ag bike' refers to all motorbikes with two, three and four wheels, used for farm work. Three and four wheelers are also known as 'all terrain vehicles' or ATVs. Threewheeled ATV's are illegal and their use and operation is strictly prohibited.

Most Ag bike injuries result from lack of training and experience, speed, uneven or unfamiliar terrain, humps, logs, rocks, embankments, carrying a passenger or an unbalanced load, inadequate protective clothing and unsafe driving. Those aged between 10 and 24 have a significantly higher risk of injury on ATVs.

Ag bike injuries are predominantly to legs, followed by injuries to spine, arms and head. Three and four wheeler spills often result in the rider being pinned beneath or rolled on by the vehicle. Assess all use of Ag bikes for likelihood and possible severity of injuries. Develop safe use procedures to match the risk.

Make the Changes

The following suggestions will help minimize risks.

- Never ride an Ag bike without an approved helmet.
- Long sleeves and pants, sturdy boots and gloves all provide protection if you come off the vehicle.
- Eye protection prevents serious eye injuries from bugs, branches or stones.

Maintenance

- Check your bike before riding it.
- Pay attention to maintenance advice in the vehicle manual.
- Check brakes and tires regularly.
- Ensure all parts are genuine or are at least equivalent components designed for use on your particular brand of bike.

Attachments

- Take extra care when using attachments such as spray tanks and other equipment on your ATV, as they can change the vehicle's center of gravity and affect its stability.
- Ensure any attachments are designed for use on your ATV.

Terrain

- — Be on the lookout for potential hazards when riding. Rocks, bumps, irrigation pipes and wildlife all have the potential to cause an accident, and should be approached with caution.
- — Take extra care when operating a bike on unfamiliar or rough terrain.
- — Where possible, use familiar farm tracks.
- — Be particularly careful when turning, approaching a rise or navigating an obstacle.

If you are not sure of your ability to clear an obstacle, find another route or go back.

Paved Surfaces and Public Roads

- — Don't drive ATVs on paved or bitumen surfaces. They are not intended for use on smooth surfaces and could be difficult to control.
- — Never ride ATVs on public roads. It may be difficult to avoid a collision if other " vehicles are on the road.

Passengers and Children

- — Passengers and ATVs don't mix. ATVs are designed to be controlled by the shifting of weight around the vehicle. A passenger limits the driver's ability to do this.
- — Never allow children to operate an ATV without training and appropriate supervision.
- — Children do not always have the weight, limb size, skill and judgement to control an ATV safely.

Stunts and Speeding

- — Never attempt jumps, wheelies or other stunts on an ATV.
- — Ride at an appropriate speed for the terrain, your experience and the visibility conditions.

Drugs, Alcohol and Fatigue

— Never ride under the influence of alcohol or drugs, including prescription drugs. They may affect your balance, vision, judgment and concentration.

— Fatigue can also limit your ability to control an ATV safely. Operating an ATV is more physically demanding than driving a car. If you are travelling long distances, take frequent rest breaks.

— Ensure you are dressed comfortably and appropriately - uncomfortable clothing can make you tire more easily.

Dangers of Farm Machinery

Farm machinery uses power to do work. This creates many possible hazards for both operators and bystanders. Even though manufacturers take many steps to make machinery safe, all hazards cannot be removed. Minor and serious injuries can occur when workers are not paying close attention, taking shortcuts, ignoring warnings or failing to follow safety rules. The wide variety of warning, caution and instructional decals placed on machinery are there for your safety.

There are many different types of farm machinery, but they all have similar characteristics and hazards. Not all these hazards can be completely shielded, so farmers must use caution when operating them.

Shear Points: Shear points exist when the edges of two objects move toward or next to each other closely enough to cut relatively soft material. Cutting points happen when a single object moves forcefully or rapidly enough to cut. They can be found on many types of crop cutting equipment, such as forage harvester heads and sickle bars, and grain augers. Shear and cutting points are hazards because of their cutting force. They often move so rapidly that they may not be visible, so it is easy to forget that they are there.

Pinch Points: Pinch points exist when two objects move together, with at least one of them moving in a circle. They are common in power transmission devices, such as belt and chain drives, feed rolls and gear drives. Fingers, hands and feet can be caught directly in pinch points or they may be drawn into the pinch points by loose clothing that becomes entangled. Contact may be made by brushing against unshielded parts or by falling against them. Shields cover most of these areas to prevent accidents, but on e caught, these machines move too fast for someone caught to get out of a pinch point. Be aware of these hazards and wear clothing that cannot be caught. Never reach over or work near rotating parts. Turn off machinery to work on it and replace any missing shields.

Wrap Points: Any exposed, rotating machine component is a potential wrap point. Protruding shaft ends can also become wrap points. A cuff, sleeve, pant leg or just a thread can catch on a rotating part and result in serious injury. Entanglement with a wrap point can pull a person into the machine or wrap their clothing so tightly the person is crushed or suffocated. A person can even lose their balance and fall into other machine parts. Even a perfectly round shaft can be a hazard if there is enough pressure to hold clothing against the shaft. Shafts that are not round increase the hazard significantly. Universal joints, keys and fastening devices also can snag clothing. Be aware of potential wrap points and shield those that can be shielded. Place warnings on those that cannot be covered or paint them a bright color.

Crush Points: Crush points exist when two objects move toward each other, or when one object moves toward a stationary object. Hitching tractors to implements may create a potential crush point. Failure to block up equipment safely can result in a fatal crushing injury. Workers need to be careful so they do not get caught in crush point areas. Crushing injuries most commonly occur to fingers at the hitching point. Wait until the tractor has stopped before stepping into the hitching position. The head or chest of an operator may be crushed between the equipment and a low beam or other part of a farm building. Usually, these accidents occur when the machine is operating in reverse. Tree limbs are also potential hazards.

Free-Wheeling Parts: The heavier a revolving part, the longer it will continue to rotate after power is shut off. Rotary mower blades, baler flywheels and various other farm machinery components will continue to move after power stops. Workers must allow time for these wheels or blades to stop before approaching them. This may take as long as two and a half minutes.

Pull-In Points: Pull-in points usually occur when someone tries to remove plant material or other obstacles that have become stuck in feed rolls or other machinery parts. Always shut off the power before attempting to clear plugged equipment.

Springs: Springs are commonly used to help lift equipment, such as shock absorbers, and to keep belts tight. Springs may harbour potentially dangerous stored energy. Know what direction a spring will move and how it might affect another machine part when released, and stay out of its path.

Hydraulic Systems: Hydraulic systems store considerable energy. They lift implements, such as plows, change the position of implement components, such as a combine header or bulldozer blade, operate hydraulic motors and assist in steering and braking. Careless servicing, adjustment or replacement of parts can result in serious injury. Highpressure blasts of hydraulic oil can injure eyes or other body parts by burning or penetrating the tissue.

Leaks are a serious hazard. Never inspect hydraulic hoses with your hands because a fine jet of hydraulic fluid can pierce the skin. Get medical attention quickly, or you could lose that part of the body that was injected. Use a piece of cardboard to test the hose for leaks. Follow the instructor's manual when servicing hydraulic systems. Make certain the hydraulic pump is turned off. Lower the attached equipment to the ground and confirm that load pressure is off the system. Treat hydraulic fluid as flammable liquid. Avoid open flames and sparks if hydraulic fluid has been spilled. Being aware of these machinery hazards is the first step to prevent accidents.

References

Chatha, I.S. and Grewal, S.S., *A study of tractorisation in Punjab*, Department of Economics and Sociology, Punjab Agricultural University, Ludhiana, 1991.

Datt, S. and Sundharam, K.P.M., *Indian Economy* S. Chand & Co. Ltd., 1998.

Gajja, B.L.; Kalla, J.C. and D.L. Vyas, *Determinants of tractorisation in arid areas of Western Rajasthan*, Agricultural Situation in India, 40(4):257-260, 1985.

Kumar, N. et al., *Farm mechanization and its impact on Punjab agriculture*, paper presented in I.S.A.E. Seminar report, Khalsa College, Amritsar, 1995.

Lekhi, R.K. and Singh, Joginder, *Agricultural Economics*,Kalyani Publishers, New Delhi, July 2002.

Mishra, P. and Sundram, *Some aspects of the economics of harvest combines in Punjab*, Economic & Political Weekly, 10(39), 1975.

PAU, Ludhiana, Directorate of Extension Education, *Package of Practices for Rabi Crops* in Punjab, 2002.

Pathak, A.S., *Farm mechanization and productivity*, Productivity, 23(4): 411-417, 1985.

Planning Commission, Govt. of India, *Expert committee on combine harvesters*, Report (Mimeographed), 1975.

Sethi, K., *Optimum land use pattern and normative demand for tractors in a growing economy-a case study of Ludhiana district*, unpublished M.Sc. Thesis, Punjab Agricultural University, Ludhiana, 1992.

Sharma, B.M.; Goyal, S.K. and G.P. Sharma, "Input use efficiency on tractor and bullock operated farms in Aligarh district of Uttar Pradesh", *Indian Journal of Agricultural Economics*, 47(3): 487, 1992.

Singh, B., *Demand for tractors in Punjab*, unpublished M.Sc. thesis, Punjab Agricultural University, Ludhiana, 1979.

10

Tools for Tillage and Seebed Preparation

The tillage operations, defined as mechanical manipulation of soil, are performed to achieve the desired seedbed to provide optimum environment for seed germination and plant growth. Seedbed preparation for sowing/planting of different crops is done through primary and secondary tillage operations.

Seedbed Preparation

The optimum seedbed preparation for raising upland crops, involves the following unit operations:

Loosening of Soil

This is done to achieve a desired granular soil structure for a seedbed and to allow rapid infiltration and good retention of moisture, to provide adequate air exchange capacity within the soil and to minimize resistance to root penetration and shoot growth. Local plough and blade harrow are traditional implements used for loosening of soil. These are simplest tools designed to break the topsoil and multi-passes are carried out to prepare seedbed. Mould board plough, disc plough, soil stirring plough, ridger plough, tool frames/ carriers with mouldboard plough or tillage sweeps, etc. are improved implements designed for breaking soil. Ploughs are used to break soil and invert furrow slice to control weeds, etc.

Clod Size Reduction

Clod breaking operation is required to produce a granular soil structure in the final seedbed. Tine cultivator and disc harrow are used for breaking of clods. Generally these are operated after one pass of mouldboard plough or ridger plough. Direct harrowing or cultivator operation is also performed when the fields are clean and free from plant residues of previous crop. Clod crushers, pateta harrow, etc., are very effective for clod crushing under favourable soil moisture conditions but their effect is confined to soil surface only. Power driven implements like rotavators disintegrate the clods over a wide range of soil moisture and provide uniform and fine size clods or aggregates in seedbed.

Clod Sorting

Operation of tools with narrow tines such as comb harrow and spike tooth harrow, in loosened soil, produces a sorting effect, bringing larger clods and aggregates on surface. The sorting effect increases with increasing forward inclination of tines and share width and decreasing speed and soil moisture. Large size clods on the surface are recommended because of their stability under rainfall, which helps in reducing soil erosion.

Compaction and Consolidation

Wide, backward inclined implements compact soil as well as break clods in top surface of soil. Direct compaction at seed depth can best be achieved using narrow press wheels/discs. Planking is widely used to compact the soil at the surface. It is done to reduce wind erosion, soil moisture loss and helps in better establishment of crop.

Smoothening of Seedbed

Smoothening of seedbed is required for proper operation of sowing machines, better distribution of irrigation water and quick disposal of excess rainwater. Smoothening can be best achieved by using wide backward inclined blades, such as levelling boards, floats and patela harrow with closely spaced shallow working narrow tines for collection of trashes. Wooden plank, patela harrow, are recommended for smoothening operation.

Optimum tillage to achieve maximum crop yields with minimum energy consumption should be the aim of seedbed preparation. A good seedbed is generally considered to imply finer particles and greater firmness in the vicinity of seeds. The depth up to which tillage operations disturb the soil can classify the operation as shallow, medium or deep. The depth of tillage depends on the crop and soil characteristics and also on the source of power or energy available.

In arid and semi arid areas with high average soil temperature and dry spells, there is a need to break the soil, which becomes very hard. A pointed tool like chisel or bar point are used on country plough to break soil without inverting or disturbing crop residue, in order to collect and store rain water and reduce wind erosion and evaporation losses. Under such conditions lister plough, rigid tine cultivator, duck foot sweeps and other similar equipment are useful and can be operated for one or two passes.

Under black soil regions (vertisols) of Madhya Pradesh, Maharashtra, Gujarat and Andhra Pradesh, soils dry up and develop deep cracks during hot summer weather, and hence ploughing is not very essential. Mould board ploughing may be done once in 3 to 4 years to destroy weeds and break hard pan. For such soils shallow cultivation by a blade harrow or sweep cultivator is sufficient to prepare a good seedbed, when weeds are under control. In humid areas, it is desirable to have deep tillage accompanied by soil inversion and burying of crop residues. This helps in enhancing nitrogen fixation in soil and incorporation of biomass.

In dryland areas tillage requirements are mainly linked with improved moisture intake and retention, reduced evaporation and checking of weed growth. Increased infiltration rate and higher crop yields can be achieved under dryland conditions by performing deep tillage by mould board plough. In brief, the recommendations are to do shallow or medium tillage in most conditions. For weed control and increased storage of rain water deep tillage is recommended. Summer ploughing with mould board plough, once in two to three years, to invert the soil, is also recommended.

PUDDING FOR RICE TRANSPLANTING

Puddling of soil generally refers to breaking down soil aggregates at near saturation into ultimate soil particles and is one of the common operations in low land rice fields. It is normally done after initial ploughing and allowing about 50 to 100 mm of standing water in the field. However; in low land condition the farmers often flood the field prior to ploughing and puddling to weaken the mechanical strength of the soil.

Retention of standing water on the rice field helps weed control and oxidation-reduction. Such conditions help achieving nutrient balance, and a soft soil suitable for transplanting rice seedlings. Puddling helps retain standing water in the rice field by producing fine soil particles that reduce soil porosity, thus reducing percolation losses of nutrients. Puddling is also beneficial because it controls weeds, levels the soil surface and provides a homogenized puddled tilth.

Puddling is done with an animal or tractor drawn implement (puddler) such as ploughs, comb harrow, patela puddler, ladder puddler and rotary puddler. The degree of puddling is however, dependent on the type of puddler and on intensity of puddling. Rotary puddlers generally are better than ploughs because their rotary motion continuously changes the direction of the shear stress and therefore matches the weakest fracture plane within a clod. Further the rotary puddlers tend to compact the sub soil, chop and press down organic matter and require relatively low draft as compared to ploughs.

Earth Moving Machinary

Backhoe and Dozer

This is an attachments of tractor. The dozer is mounted in front of the tractor and backhoe in the rear. The dozer and backhoe can be easily removed and attached to the tractor. The dozer consists of a thick curved plate and hardened strip. The strip has sharp cutting edge and is joined to the curved plate of the dozer with fasteners. Therefore, the strip can be replaced on wearing or becoming blunt. The dozer plate is joined to the tractor with sturdy arms and can be raised or lowered with hydraulic system of the tractor. Backhoe consists of a bucket with digging fingers, hydraulic cylinder, arms and base for attaching to the rear of the tractor. The bucket position is manipulated by hydraulic system. The digging fingers are hardened and can be replaced on wearing or becoming blunt.

Backhoe is used for excavating soil, making foundation for building, making trenches for pipe and cable laying, garbage handling, widening of rural roads and removal of bushes and trees etc. Dozer is used for agricultural land levelling, making bunds in the farms and terracing of farm, road making and site clearance, for trench filling at dam project, after laying cable or pipe etc.

Backhoe and Loader

This is also an attachments of tractor. The loader is mounted in front of the tractor and backhoe at the rear. The loader and backhoe can be easily removed and attached to the tractor. Loader consists of curved frame in the front, which acts as bucket, arms hinged to each other by pins, hydraulic cylinders and hydraulic system. The loader, therefore, can be maneuvered to the required position through hydraulic controls.

Loader is used for removal of mud and loose soil at canal worksite, for loading of crushed stones into the dumpers and trailers at stone crusher unit, for loading of salt into trucks and trailers, for handling of clay and soil and loading it into the trucks and trailers etc.

Tractor Mounted Blade

It is attached to the tractor with the 3-point linkage system and is hydraulically controlled. It consists of replaceable blade attached to curved steel body, side wings and indexing arrangement for tilting and angling of the blade. The cutting blade is made of medium carbon or alloy steel and the cutting edge is hardened and sharpened. The blade with body is also called mould board. The mouldboard can be angled left or right by lifting the spring Joadedlatch pin and by turning the mouldboard. To tilt the blade for ditching or terracing the blade is tilted to the desired angle by moving the index pin. The depth of cutting is controlled by hydraulic system of the tractor. The blade can be pitched forward and back or tilted at 15° to 30° left or right. It can be reversed for back filling. For increasing the length of blade, extensions are provided. The tracer blade is used for grading, levelling of fields, filling of depressions and smoothening of field for irrigation.

Hydro-dozer

This is a tractor front mounted dozer. The dozing blade assembly consists of a curved plate (bucket) to which a cutting blade made of hardened alloy steel is joined with dozer arms, hydraulic cylinders, frame and body. The cutting blade edge is sharp and fastened to the plate with fasteners. The blade is reversible type. The dozer is hydraulically controlled. For land shaping, field terracing, construction of roads, contour bunding etc.

Hydraulic Scraper

The hydraulic scraper is towed behind the tractor. The scraper consists of cutting blade, hydraulic system, hitch point, hitch bar, apron, bowl, wheels, apron cylinder, side frame? bucket cylinder, spring, and side arm. The scraper working is controlled by the hydraulic arrangement. The blade is made of alloy steel and has self-sharpening tungsten carbide cutting edge. For operation, the scraper is attached to the tractor, hydraulic system connected and apron is raised. With the forward movement of the tractor, the blade penetrates into the soil and fill bucket bowl gets filled. The apron is closed after the bucket is filled and the scraper is moved to the point of unloading. For unloading, the bucket is tilted hydraulically. It is used for collecting the soil from one place and unloading at the other. It is used for rough leveling, cutting of high spots and filling of depressions.

Sheep Foot Roller

The sheep foot roller consists of box frame of heavy angles, hard faced solid forged tamping feet, heavy-duty shaft enclosed in the drum, steel drum on which forgetted tamping foot are welded, reinforced box section draw bar provided with heavy duty swivel hook and adjustable cleaner teeth to prevent accumulation of dirt between the teeth. The roller is mounted on heavy-duty frame. The equipment is available in the form of one or two drums and is attached to the drawbar of the tractor. Independent oscillating frame and oscillation stopper are provided in the double drum sheep foot rollers. It is used for compaction of soil and farm roads.

Culti-packer

The implement consists of number of cast iron V shaped rollers mounted on three axles. The implement can be operated either with one or 3 gangs. Some times it can be attached behind disc harrow (trailing type) to affect compaction for moisture conservation. Breaking of clods and packing of soil thereby conserving moisture.

Vibratory Roller

The vibratory roller consists of frame, which is a welded structure and carry the engine at the rear in static balance condition, roller fabricated from steel plate with adjustable scrapers fitted in front and rear, vibratory drive having spring loaded centrifugal clutch directly mounted to the flywheel, vibrating mechanism and towing mechanism. The vibratory roller is towed with the tractor for operation. Vibratory mechanism provided in the roller helps in achieving higher compaction. It is used for compacting filled-in earthwork such as embankments, sub-grade, sub-bases for farm roads, building foundation etc.

Ditcher

It consists of two curved wings with cutting blades, front cutting point, tie bars for adjusting wingspan, and hitch assembly with 3-point linkages. The cutting blades and cutting point are made of medium carbon or alloy steel, hardened and sharpened. The ditcher is operated by tractor and controlled by hydraulic system. The ditcher penetrates in the soil due to its own weight and suction of the cutting point. Upon drawing the ditcher in the field it opens the soil in the shape of ditch with either 'V' bottom or flat bottom. The wings enable the ditcher to slice and roll the tough sod, brush and root sets. The depth and width of the ditch is adjusted from the operators seat. The front cutting point and wings cutting edges are replaceable. It is used for making ditches for irrigation and drainage.

Rotary Ditcher

It consists of a rotary cutter operated by pto shaft of the tractor, gear box, 3-point linkage, hitch system, frame, body, deflector and ditch former. The machine is operated by tractor. Rotary cutter is main component of the ditcher and it consists of drum fitted with cutting knives or cutters. The knives are of different shapes and are made of alloy steel by forging. The knives can be replaced on becoming blunt. The rotary cutter excavates soil, which is uniformly distributed to one side. The deflection of the soil can be adjusted by the deflector. Ditch former, having trapezoidal shape fitted in the rear, form the ditch. It is used for making ditches for irrigation and drainage.

Improved Bakhar

The improved bakhar consists of a frame made of—mild steel angle section in which a V -shaped blade is attached by means of fasteners. Behind the blade a roller made of 150rnm diameter Mild Steel pipe is provided to crush soil clods. The frame also supports a handle by which the operator controls the implement. The unit is suitable for seedbed preparation in black soil. This has a modified V-blade replacing conventional curved blade for reduced draft. A pair of bullocks is used for pulling the implement. It is used for seedbed preparation in black soil.

Animal Drawn Mouldboard Plough

The implement consists of a share, mouldboard, landside, handle, depth gauge wheel and wooden beam. The share is made from medium or high carbon steel and hardened to resist wear during operation. The mouldboard that inverts the soil is made of mild steel. The gauge wheel is used to control the depth of operation. It has convenient handles for easy operation. For operation the plough is moved in the soil with the help of a pair of bullock. Share of the plough penetrates into the soil and makes a cut below the soil surface. Mould board lifts, pulverizes and inverts the furrow slice. Animal drawn mouldboard plough is used for ploughing and inversion of soil.

Melur Plough

This type of plough is used in Madurai, Ramnad and Tanjore districts of Tamil Nadu. This plough is suitable for ploughing in wetlands for raising rice crop using a pair of bullocks. The depth of cut in black soil is 3.5 cm and the width is 100-130 mm. The shoe is made of single casting with ribbed surface. The pole shaft is made of babul or vengai wood. This type of plough is used for seed bed preparation and puddling in wetlands.

Improved Iron Plough

The bullock drawn improved iron plough is made of mild steel except the pole shaft and hence it has longer life. As and when the share wears off, it can be pushed forward. Pole shaft angle and height of handle can be adjusted according to field requirements. It is operated with one pair of bullocks. The plough is provided with a mouldboard as an optional attachment for soil inversion. This implement could be used for dry ploughing operations in all types of soil. The plough is provided with a mould board as an optional attachment for soil inversion.

Animal Drawn Bose Plough

It is a mould board type plough suitable for upland paddy and dryland cultivation. It consists of wooden beam, Wooden handle with a grip, wooden body and a share cum mould board made of mild steel. The share cum mouldboard has sharp point and cutting edge, which cut and invert the furrow slice. Upon becoming dull it can be replaced. It is suitable for upland paddy and dryland cultivation.

Local Plough

Khargaon Type Hal

The plough consists of mild steel sheet body to which a replaceable share is attached. The body also serve the purpose of frame and the other component like handle, tie bar and hitch assembly is also attached to it. The shape of the share is similar to flat iron section with a tang and is held on the body with fasteners. The share is usually fabricated from used leaf spring steel; tip is forged to a sharp edge and hardened. The share can be replaced after it has worn out and the cutting edge can be sharpened upon becoming dull. The operation of the plough is similar to other bullock drawn ploughs. The hitching point for the beam is a swing type mechanism, which is adjusted by the help of the brackets using the pre-drilled holes and attached to the main body of the plough. It is used for primary tillage.

Dabra Plough

The complete plough is made of steel. It consists of mild steel plate frame to which other parts such as plough sole; handle and soil working components are attached. The share of the plough is fabricated from mild steel sheet having a triangular shape. Another piece of sheet is joined to the share which acts as mouldboard. The share end and mouldboard assembly is 15 tilted slightly for penetration and inversion of the furrow slice. The share upon becoming dull or after it is worn out can be replaced. The plough is suitable for tillage in light soils. It is used for primary tillage in light soils.

Rau Type Hal

The plough consists of triangular shape body made of mild steel sheet, handle body made of mild steel sheet, share made of used leaf spring steel, hitch assembly and tie bars made of mild steel flat iron sections. The share is forged to a screwdriver point shape with tang. The share is held to the triangular body of the plough with the help of U -clamps. The share after becoming dull can be sharpened or replaced. Similar to other ploughs found in the region, it has special brackets having a number of holes to which the beam can be attached. These holes serve the purpose of adjusting the plough suiting to the height of the animals being used. The handle is also placed at a convenient height for the operator. This implement is popular in the Malwa regions of Madhya Pradesh. It is used for primary tillage in kharif and rabi seasons.

Animal Drawn Chisel Plough

It is a pointed curved bar type implement useful for breaking hard layer of soil below the normal ploughing depth to facilitate infiltration of rainwater. Use of this implement leads to increase in yield by up to 15 per cent compared to conventional method of field preparation by country plough without any deep ploughing. It is usually operated by a pair of bullocks. The equipment is commercially available. It is a pointed curved bar type implement useful for breaking hard layer of soil below the normal ploughing depth to facilitate infiltration of rainwater.

Mp Iron Wedge Plough

The plough is similar to the indigenous plough in shape. It consists of mild steel box section frame, a shank attached to the frame, plough body made of mild steel sheet attached to the shank, share attached to the body of the plough, hitch assembly and handle. The share point is fabricated from used leaf spring steel and forged to the required shape. The share can be replaced upon becoming dull. It requires a pair of bullocks to pull; the share cuts a slice, which is inverted on both sides by the body of the plough. It has several holes on the brackets meant for attaching the beam such that the implement can be easily adjusted for varying heights of animals. The handle provided at a convenient height enables the operator to guide it with ease when pulled by the animals.

Birsa Animal Drawn Ridger Plough

It is a ridger plough to perform all functions of country plough with added efficiency and advantages. It is suitable for hilly region of Chota Nagpur region. It saves 9 per cent labour and operation time and 5 per cent cost of operation compared to the conventional method of using country plough. The Birsa ridger plough was found to be better than the country plough in its performance. It performs all functions of country plough with added efficiency and advantages.

Animal Drawn Kapas Ridger

The ridger is more popular in the cultivation of cotton, therefore has been named as kapas ridger. The ridger consists of a sturdy frame, share, two mouldboards with tail, head wheel, double handle and hitch assembly. The share of the ridger is made from medium carbon steel or alloy steel and hardened to resist wear in the soil. The mouldboards are adjustable and the wingspan can be adjusted according to row width. Tailpieces are provided for increasing the furrow width. The head wheel regulates the depth of the furrow. For operation, the implement is drawn by a pair of bullocks, the depth of furrow is 18 adjusted by the head wheel, width of the furrow is adjusted by the wingspan of mouldboard and tail pieces. The ridger has two handles, which helps in keeping the implement stable during operation. The furrow is formed in a single pass and the ridge in the return pass. It is used for making ridges and furrows for cultivation in maize, sorghum, cotton, millets, cowpea, groundnut and other arable crops.

Bullock Drawn Ridger

The ridger plough consists of single share and double mould board attached to a short (steel) or long beam (wooden). Two handles are provided to balance the ridger during operation. During operation the share cuts furrow slice and pulverises the soil, which moves on to the mouldboard and is inyerted. During the onward movement, a furrow is formed and ridges get formed in its return pass. The point of the ridger can be replaced after it is worn out. In certain designs, the width of ridges can also be adjusted. It is used for making ridges and furrows for potato and sugarcane planting and making field channels for irrigation.

Bullock Drawn Disc Harrow

The harrow consists of a frame and shafts on which discs are mounted. The shafts are held in bearings for smooth rotation of discs. The shafts are called gangs on which discs are arranged and equally spaced with the help of spoos. Single gang and double gang disc harrows are equally popular and operated using a pair of animals. The discs are fabricated from carbon steels and the cutting edges are hardened. The harrow is provided with an operator's seat and also has a transport wheel which aids in easy transportation. The operator's seat enables the operator to ride instead of walking, which helps in deeper penetration and reduces drudgery. It is used for seedbed preparation, cutting and burying of grass and weeds.

Animal Drawn Patela Harrow

The animal drawn patela harrow consists of a heavy batten made of Sal wood having 1500mm length and 100mm thickness on which a Mild Steel angle frame is fixed by means of screws. The frame carries a bar to which curved and pointed hooks are attached. The bar can be raised or lowered by means of the lever having a slotted sector to lock its position. Some models may be up to 2000mm in length. It is secondary tillage equipment for clod crushing, stubble or trash collection, levelling and smoothing of land surface before seeding. It is a secondary tillage equipment and used for breaking of clods, collection of stubble or trash and levelling of the field surface before the application of seedlings.

Animal Drawn Disc Harrow

Blade Harrow

The implement consists of a blade, blade holding prongs, wooden body and a wooden beam. The blade is attached to the bakhizr frame prongs with the help of rings, which are hammered 21 into place. It cuts a thin slice of soil and removes the weeds etc and requires a pair of bullocks to pull the implement. The V-shape of the blade has a self-cleaning advantage, which prevents adherence of weeds and its roots, thus enabling speedier operation and improving the uniformity of cut as compared to the straight bladed version. It is used for the primary tillage operations.

MP Single Dora

The implement consists of mild steel box type frame, a weeding blade, blade holding prongs, hitching arrangement, handle and a steel beam. All the parts of the dora are made of steel. The blade is attached to the dora prongs with the help of rings. The bracket for fixing the beam has multiple holes so that the height of the beam can easily be adjusted according to the height of the animals being used. The implement is pulled by a pair of bullocks and scraps the soil. The shape of the blade is slightly concave. This implement has multiple utilities and can also be used for exposing potato tubers and digging out groundnut vines from the soil. Once its soil-working component gets worn out it can be reshaped and fitted again to improve its performance. It is used for the weeding of crops sown in line such as cotton, maize, sorghum and similar crops and also for digging out potato and groundnut.

MP Double Dora

The implement consists of two mild steel box type frame, two weeding blades, blade holding prongs for each blade, common hitching arrangement, flat iron curved part to join the box frames, handle and a steel beam. All the parts of the dora are made of steel. The blade is attached to the dora prongs with the help of rings. The bracket for fixing the beam has multiple holes so that the height of the beam can easily be adjusted according to the height of the animals being used. The implement is pulled by a pair of bullocks and scraps the soil. The shape of the blade is slightly concave. With the dora, two rows can be weeded in a single pass, thus improving its field capacity but requires a heavier pair of animals. It is used for the weeding of crops sown in line such as cotton, maize, sorghum and similar crops.

Bullock Drawn Bund Former

The implement consists of two blades, flat iron frame bent at an angle a handle attached to the frame with tie bars and wooden beam. The operator's handle is made of wood for providing better grip and convenience in operation and attached to the frame with the help of suitable brackets. The frame is bent at an angle and has holes for adjusting the space between the blades. The profile of blades is made to a shape so that bund formed is trapezoidal and remains stable. The blades are attached to the frame with fasteners. For operation, a pair of bullock pulls the implement, the blades gather the loose soil and accumulate it in the form of bund.

Patela-puddler

The unit consists of a frame made of mild steel frame, a wooden plank, a gang consisting of pegs for turning the soil, and sliding type pegs (10 nos.), handle and hitch system. The operator can stand on the plank while in operation to provide additional depth. A pair of bullocks is used for the operation of the implement to prepare homogenized puddled tilth for mechanised paddy transplanting. It breaks the soil clods to obtain a smooth puddle. Soil moisture is maintained near the saturation level at the time of operation. It is suitable—for shallow puddling with high mechanical dispersion of soil particles. Used for shallow puddling of the field surface and higher mechanical dispersion of soil.

Bullock Drawn Puddler

Various designs of puddler are available and their shape, size and other features differ from region to region. Basically, the unit consists of frame, steel blades mounted on a shaft in a staggered fashion either on straight or inclined position, and operator's handle and hitch 24 arrangement. The shaft carrying blades rotate in the bushes mounted on the frame. A pair of bullocks pulls the puddler. The blades on the implement rotate and impart a lateral and turning action on the soil particles, thus achieving a good puddle. It is used for puddling of soil for paddy cultivation.

Animal Drawn Puddler

It is a rectangular blade type puddler suitable for puddling operation under wetland conditions. It has a wooden frame on which bushes are mounted using mud steel flats. The shaft carrying the blades fixed in a staggered fashion rotates in the bushes. The blades are fixed at an angle to the direction of motion. The blades on the implement rotate and impart a lateral and turning action on the soil particles, thus achieving a good puddle. It has an operator's seat, which helps in continuous operation and reduces drudgery. It saves 66 to 88 per cent labour and operating time and 66 to 82 per cent on cost of operation compared to conventional method by using bullock drawn country plough. The implement is very popular among the farmers of Andhra Pradesh. It is used for puddling operation under wet land condition.

Animal Drawn Helical Blade Puddler

It is a helical blade type of puddler suitable for puddling in wetland conditions and also for cutting and mixing of green manure crop. The implement consists of tow angle iron brackets carrying a bush. A shaft carrying specially shaped slots accommodates the blades. These blades are imparted a twist along their length to form a helix, thus its nomenclature. The blades on the implement rotate and impart a lateral and turning action on the soil particles, thus achieving a good puddle. The handle for the operator is made of wood and helps in comfortable operation. It saves 30 per cent labour, 46 per cent operating time and 30 per cent on cost of operation compared to conventional method of using country plough or tractor with cage wheels for puddling. It is suitable for puddling wetland and also for cutting and mixing of green manure crops.

Bullock Drawn Land Leveller

The bullock drawn land leveller is fabricated from steel sheets and two handles are fixed to it with the help of fasteners. Mils steel round bar is bent to a suitable shape with holes at its extremities for fixing to the scraper. This steel round has a semicircular loop to which the rope from the yoke of the animal can be attached for towing. One pair of animals is required to operate the implement. The levellers are of different design and shape, but the scoop type and tracer blade type are most common. The leveller is used for filling minor depression and levelling minor irregularities in the field. Scoop type has got a bucket of 0.15-0.2 cu.m (approx.) and is operated by a person walking behind it. The tracer blade type has got sitting arrangement and is provided with pneumatic wheels. Initial ploughing and loosening of the soil is necessary before the leveller could be put to use. It is used to fill minor depressions and to remove minor irregularities of the field.

Bullock Drawn Cultivator

The bullock drawn cultivators are available in fixed or expandable frame designs. The unit consists of reversible tines, frame, handle and a wooden beam. The working width of the cultivator can be adjusted by varying the spacing between the tines or expanding the frame. For operation the implement is attached to a pair of bullock by beam and moved in the soil. The beam is attached to the unit with the help of V-bolts and can be adjusted as per the height of the animal pair. The operator's handle can also be adjusted by varying the position of the supporting bracket. The tines dig into the soil and cut a small furrow slice. During weeding operation, the tines also uproot weeds and create soil mulch. Seeding attachment can also be provided on the frame, increasing the versatility of the unit. It is used for shallow ploughing, weeding and intercultural operations.

Tractop Mounted Mouldboard Plough

It is a tractor-operated implement and consists of share point, share, mouldboard, landslide, frog, shank, frame and hitch system. The share point is of bar type and is made from high carbon steel or low alloy steel. The share is also made from high carbon steel or low alloy steel. Both are hardened and tempered to suitable hardness (about 45 ARC). The working of the plough is controlled by hydraulic system lever and three-point linkage. Its bar point makes it suitable to break hardpan of the soil.

Ploughs are used for primary tillage operations. It cuts trash and buries it completely. It is also used for turning green manure crop for decaying under the soil, which adds humus to the soil. Compost, farmyard manure or lime when spread in the field, the mould board plough is used for turning and mixing these materials in the soil.

Tractor Drawn Disc Plough

The plough consists of common mainframe, disc beam assemblies, rockshaft category -lor category -2, a heavy spring loaded furrow wheel and a gauge wheel. In some models disc plough is designed to operate as 2, 3 or 4 bottom, by adding or removing the sub beam assemblies according to requirement. The disc angle ranges from 40 to 450 to obtain the desired width of cut and the tilt angle ranges from 15 to 250 for penetration. The discs of the plough are made of high carbon steel or alloy steel and the edges are hardened and sharpened. The discs are mounted on tapered roller bearings. Scrapers prevent soil build up on the discs in sticky soils. The furrow slice rides along the curvature and is pulverised to some extent before being thrown. Disc plough is used for primary tillage and is especially useful in hard and dry, trashy, stony or stumpy land conditions and in soil where scouring is a major problem.

Tractor Mounted Disc Harrow

The tractor mounted disc harrow consists of two gangs of discs mounted one behind the other. The discs on the front gang throw soil outward and the rear gang inward. Therefore, no soil remains uncut by the offset disc harrow. The harrow consists of a sturdy frame, discs arranged in two gangs, and three-point linkage. Discs are important component of the harrow and are made from high carbon steel or alloy steel; the cutting edges are hardened and tempered to suitable hardness. The gangs can be moved in either direction on the hitch frame. The rear gang can be moved the same amount as the front gang. When operating in orchards or plantations, the harrow can be offset to the right or left, thus enabling soil to be thrown towards or away from the trees. The offset -feature makes it possible to work under low-hanging branches. Discs with notches on the outer rim are also available for c~ operation in weed-infested fields. Three point hydraulic linkage and hydraulic control makes it highly manoeuvrable. The disc harrow is used for primary and secondary tillage. It is ideal for field disking, especially in orchards, plantations and vineyard. It is suitable for working under trees close to bunds and fence posts.

Springtyne Cultivator

Cultivator consists of a frame, tynes with reversible shovels, land wheel, hitch system and heavy-duty springs. The function of springs is to save the cultivator tynes from breaking when some hard object comes in contact with the shovel or under the tyne. The shovels are made of heat-treated steel for longer life. The implement is mounted type and is controlled by the hydraulic system of the tractor. Cultivators are used for seedbed preparation both in dry and wet soils. It is also used for interculture purpose by adjusting the tynes in wider row crops. It is also used for puddling purposes.

Rigidtyne Cultivator

It is a tractor-mounted implement and consists of mainframe made of box section, rigid tines, Uclamps and shovels. The clamping of tines makes possible to adjust the distance between them according to crop rows. The shovels are made from medium carbon steel or low alloy steel, hardened and tempered to suitable hardness. The shovels are mounted on the tynes with fasteners and can be replaced easily on wearing or, becoming dull. The depth of operation is controlled by the hydraulic system of the tractor. The shovels can be replaced by duck foot sweeps for shallow tillage.

A versatile implement used for loosening and aerating the soil and preparing seed beds quickly and economically. It is useful for subsoil cultivation and also eliminates the use of plough even for hard soils. The duck foot shovel version is ideal for surface cultivation i.e. shallow cultivation and interculture.

Rigid Tynes-hovel Type Cutivator

These are one of the most selling cultivators used in agricultural operation. It consists of a rectangular frame made of mild steel angle or channel section, heavy-duty rigid tynes made of mild steel flat or plate section, reversible shovels joined to tynes with fasteners, and hitch assembly. The tynes are mounted on the frame and distance between them can be varied. The shovels are made of medium carbon steel, old leaf spring steel or low alloy steel and hardened to 40-45 HRC. It is used for loosening and aerating of the soil, preparing seedbed, subsoil cultivation, and weeding / interculture.

Bar-point Cultivator

The cultivator consists of rectangular frame made of mild steel angle or channel section. Rigid tynes of heavy flat section are mounted on the frame with the help of clamps. Bar point shares are joined to the end of tynes. Bar points shares are made of medium carbon or low alloy steel, hardened and tempered to suitable hardness. Bar points can be replaced OR wearing or becoming dull. In place of bar points, duckfoot shovels can be attached to the cultivator. The cultivator is mounted type and is operated by tractor. The tynes of cultivator can be adjusted vertically by moving in the clamp. The cultivator is used for inter cultivation in light and heavy soils for loosening and aerating the soil, preparing of seedbeds, and subsoil cultivation. The duck foot shovel version is ideal for shallow cultivation and interculture.

Duck-foot Cultivator

The duckfoot cultivator consists of a box type steel rectangular frame, rigid tines and sweeps. The sweeps are triangular in shape similar to foot of duck, hence called duckfoot cultivator. The sweeps are made from old leaf spring steel and joined to tines with fasteners, which makes them replaceable after being worn out or becoming dull. The tines are made of mild steel flat and forged to shape. It is a tractor-mounted implement and depth of operation is controlled by hydraulic system. The cultivator is popular in black cotton soils. Sweeps are attached to these tines. Three-point linkage of the tractor is attached to the implement. The implement is mostly used for shallow ploughing and in hard soils. Used for primary tillage operation, destruction of weeds and retention of soil moisture.

Tractor Drawn Basin Lister

It is an attachment fitted behind 9-tine tractor mounted cultivator. It is used for water harvesting and conservation of soil and moisture. The lister attachment consists of mild steel framework, ground wheels with lugs, lister bottoms (trencher), roller follower, cam, hitch bracket and reversible shovel. The cultivator tills the soil and the lister bottoms form the trenches. It is used for water harvesting and conservation of soil and moisture.

Tractor Drawn Spiked Clod Crusher

It is used as a combination tillage tool with tractor drawn harrow or cultivator. The clod crusher consists of a mild steel sheet drum and pegs are welded on its surface, a rectangular frame made from mild steel angle section, hitch frame and a shaft for carrying the drum. The shaft is mounted on bearing pedestals. The soil clods are pierced and broken by the pegs of the crusher. It is used as a combination tillage tool with tractor drawn harrow or cultivator. It is suitable for breaking and segregation of clods for seedbed preparation after paddy harvest.

Peg Type Puddler

The peg type puddler is a tractor-mounted implement using the three-point linkage of a tractor of 35 hp or more. It consists of a frame made of mild steel angle sections on which the three cross bars and the cleats for the three-point linkage are welded in place. It is operated when the soil moisture is near the saturation level to obtain a fine tilth and good puddle to facilitate mechanical transplanting. For achieving better performance, the tractor is fitted with cage wheels to improve traction and achieve higher field capacity. Used to break the soil clods and puddle the soil for rice transplanting.

Pulverising Roller

Pulverising roller is an attachment to commercially available cultivator. The pulverizing roller consists of star wheels, central shaft, pulverising members, mounting link and depth controlling tee. The pulverising members are similar to lawn mower blades and are inserted in the cast star wheels in such a way that it forms helical shape and progressively come in contact with soil. The roller is attached to the cultivator with the help of two mounting links having bearing housing on one side and tensile springs on other side. Pulverising rollers are used for puddling as well as dry seedbed preparation in two runs, and creates good puddle. It saves 20-35 % fuel cons-umption and 20-30 % water requirement in comparison to traditional method.

Tractor Drawn Combined Tillage Tool

It is a tillage tool having cutting blade with rotary slicer making it suitable for seedbed preparation in one pass under irrigated soil conditions. It consists of a frame, hitching bracket, gearbox, chain drive system, cutting blades, rotary slicer and cover. The blade mounted in front loosens the soil. The rotary slicer operated with tractor pto pulverises the loosened soil. The tractor drawn combine tillage tool can serve the purpose of mouldboard plough and cultivator. It is suitable for seedbed preparation in one pass under irrigated soil conditions.

Land Leveller

It is a tractor-mounted implement controlled by tractor hydraulics and three-point linkage. It consists of hitch system, replaceable cutting blade with sharp edge, and a curved plate with side wings, which form a bucket. The blade is made from medium carbon steel or low alloy steel, hardened and temp-ered to suitable hardness. During operation, the blade digs into the soil and extra soil is collected in the bucket, which is released in the depressions of the field. The angle and pitch of leveller is adjustable. The leveller can also angled left or right, or reversed for back filling. It is suitable for land prepar-ation operations such as scraping, grading, levelling and back filling. It is also used for irrigation, terrace work and general cleaning of field.

Tractor Drawn Leveller

The leveller consists of frame, 3-point linkage, cutting or scraping blade, and thick curved sheet closed from sides to form a bucket. The scraping blade is made from medium carbon steel or low alloy steel, hardened and tempered to about 42 HRC. The blade is joined to the curved sheet with fasteners and can be replaced after being worn out or becoming dull. The working depth of the implement is controlled by hydraulic system of the tractor. Levelling of fields and pulling or pushing loosened soil from one place to other.

Channel Former

The channel former consists of two inner blades, two outer blades, hitch frame, mainframe and shovel. The front portions of the two inner blades are joined together and form an angle of 300 in between them. At the junction of these two inner blades a cultivator shovel is fixed to penetrate into the soil. The inner blades can be mounted 50 to 100 mm lower than the outer blades and form a furrow at a lower depth than the surface of the bed for the flow of irrigation water. The two outer blades are placed one on each side of the inner blades and at an angle of 600 to the direction of the travel. The soil collected from the furrow is formed as bund on both the sides of the irrigation furrow. It is used for making channels and beds at regular intervals for irrigation.

Tractor Bund Former

The bund former consists of mild steel angle iron frame; hitch system, and two blades (wings). The blades are made by mounting mild steel sheet on an angle iron frame. The blades are adjusted in converging manner and has wider opening in the front in comparison at the rear end. The distance between blades can be adjusted according to size of bund required. The implement is mounted type and operated in tilled soil. Making of bunds in the field.

Tractor Drawn Channel Cum Bund Former

It is a tractor-mounted implement and consists of inner wing, adjustable handle, frame, hitch pin and outer wing. The wings are reinforced or tensioned at, the bottom edge and are made from thick mild steel sheet. It is operated under well-prepared and pulverised soil condition and can be used under all types of soil and crop conditions. The distance between wings can be adjusted according to the size of bund. It is used for making field channels to manage the irrigation water effectively and making small bunds across the slope for inter-plot rainwater harvesting under rainfed areas.

Tractor Drawn Ridger

The ridger is used in sugarcane growing area of the country. It consists of rectangular frame made of mild steel angle or channel section, 3-point hitch assembly, shanks and ridger body. The ridger body consists of two mouldboards, share, point and tie bars to vary the wingspan of ridgers. The share point is made from medium carbon steel or low alloy steel, hardened and tempered to about 42 HRC. Upon wearing or becoming dull the share point can be replaced. The ridger is operated in tilled soil by a tractor, the are point penetrates in the soil, ridger body displaces the soil to both sides and a furrow is created. The soil mass between furrows forms a ridge. The depth of operation is controlled by hydraulic system of the tractor. It is used for making furrows and ridges for sugarcane, cotton, potato and other row crops.

Tractor Drawn Bed-Furrow Former

The tractor drawn bed-furrow former consists of mild steel angle iron frame; three point linkage, lifting pin, furrow former, bed former and stiffener. The bed and furrow formers are made of mild steel sheet and bent in required shape. The stiffener are used to strengthen the fortners. The implement is operated in the tilled soil. It is used for forming alternate beds and channels. The beds are suitable for planting crops like sorghum, maize, cotton. This bed and furrow system is ideal for efficient irrigation management.

Tractor Mounted Subsoiler

It consists of beam made of high carbon steel, beam supports which are flanged at upper and lower edges for rigidity, hollow steel adaptor welded to bottom end of the beam to accommodate share base, share base having square section, share plate made from high carbon steel and shank drilled and counter bored for set board which secures the base in the adaptor. Share plate is made from high carbon steel, hardened and tempered to suitable hardness. Two symmetrically located bolt holes allow reversibility of share. The working depth of the subsoiler is controlled by hydraulic system and linkage of tractor. It is used to break hardpan of the soil, loosening of the soil and helps the water to seep into the soil for improving drainage. A mole ball can be attached to create a small tunnel in the soil, which serves as drainage channel for water.

References

Kumar, N. et al., *Farm mechanization and its impact on Punjab agriculture*, paper presented in I.S.A.E. Seminar report, Khalsa College, Amritsar, 1995.

11

Harvesting Tools and Equipment

Crops are harvested after normal maturity with the objective to take out grain, straw, tubers etc. without much loss. It involves cutting/digging/picking, laying, gathering, curing, transport and stacking of the crop.

In case of cereals like wheat and paddy the plants are straight and smooth and ears containing grains are at the top whereas most of oilseed and pulse crops have branches, which create problems' in harvesting by manual or mechanical means. As per Bureau of Indian Standards the cutting and conveying losses should not be more than 2 per cent.

The harvesting of crops is traditionally done by manual methods. Harvesting of major cereals, pulse and oilseed crops are done by using sickle whereas tuber crops are harvested by country plough or spade.

All these traditional methods involve drudgery and consume long time. Timeliness of harvest is of prime importance. During harvesting season, often rains and storms occur causing considerable damage to standing crops. Rapid harvest facilitates extra days for land preparation and early planting of the next crop.

The use of machines can help to harvest at proper stage of crop maturity and reduce drudgery and operation time. Considering these, improved harvesting tools, equipment, combines are being accepted by the farmers.

Type of Mechanical Harvesting Tools and Equipment

Serrated Sickle

It has a serrated curved blade and a wooden handle. The handle of improved sickle has a bend at the rear for better grip and to avoid hand injury during operation. Serrated blade sickles cut the crop by principle of friction cutting like in saw blade. The crop is held in one hand and the sickle is pulled along an arc for cutting. Cutting of crop close to the ground is possible with modified handle. Energy requirement is 80-110 man-h/ha. It can be used effectively for harvesting of wheat, rice and grasses.

Reapers

Reapers are used for harvesting of crops mostly at ground level. It consists of crop-row divider, cutter bar assembly, feeding and conveying devices. Reapers are classified on the basis of conveying of crops as given below:

- *Vertical conveying reaper windrower:* It consists of crop row divider, star wheel, cutter bar, and a pair of lugged canvas conveyor belts. This type of machines cut the crops and conveys vertically to one end and windrows the crops on the ground uniformly. Collection of crop for making bundles is easy and it is done manually. Self-propelled walking type, self-propelled riding type and tractor mounted type reaper-windrowers are available. These types of reapers are suitable for crops like wheat and rice. The field capacities of these machines vary from 0.20-0.40 ha/h.
- *Horizontal conveying reapers:* This type of reapers are provided with crop dividers at the end, crop gathering reel, cutter bar and horizontal conveyor belt. They cut the crop, convey the crop horizontally to one end and drop it to the ground in head-tail fashion. Collection of crop for making bundles is difficult. This type of reapers are tractor mounted and suitable for wheat, rice, soybean, and gram. Performance of reapers with narrow-pitch cutter bar is better for soybean and gram crops.

— *Bunch conveying reapers:* This type of reapers are similar to horizontal conveying reapers except that the cut crop is collected on a platform and is released occasionally to the ground in the form of a bunch by actuating a hand lever. Collection of crops for making bundles is difficult. Bullock drawn and tractor-operated models are available and they are suitable for harvesting wheat, rice and soybean crops.

— *Reaper binders:* The cutting unit of this type of reapers may be disc type or cutter bar type. After cutting, the crop is conveyed vertically to the binding mechanism and released to the ground in the form of bundles. Self-propelled walking type models are available but these are not popular due to high cost of twine. Reaper binders are suitable for rice and wheat.

Strippers

The design of a tractor front mounted stripper is available for collection of matured grass seeds from the seed crops. It consists of a reel having helical rubber bats which beat the grass over a sweeping surface where the ripened seeds get detached and the seeds are collected in the seed box.

Diggers

The design of groundnut and potato diggers of animal drawn and tractor operated types are available. The digging units consists of V -shaped or straight blade and lifter rods are attached behind the share. These lifter rods are spaced to allow the clods and residual material to drop while operating the implement. The plant along with pods/tubers is collected manually.

Combines

Various designs of combine harvester having 2 to 6 m long cutter bar are commercially available. Therefore, no research effort has been initiated till now. However, the need of a small whole crop combine harvester is felt. The function of a combine harvester is to cut, thresh, winnow and clean grain/seed. It consists of header unit, threshing unit, separation unit, cleaning unit and grain collection unit. The function of the header is to cut and gather the crop and deliver it to the threshing cylinder.

The reel pushes the straw back on to the platform while the cutter bar cuts it. The crops are threshed between cylinder and concave due to impact and rubbing action. The threshed material is shaken and tossed back by the straw rack so that the grain moves and falls through the openings in the rack onto the cleaning shoe while the straw is discharged at the rear. The cleaning mechanism consists of two sieves and a fan. The grain is conveyed with a conveyor and collected in a grain tank.

Use of Mechanical Harvesting Equipment

The following criteria are must for successful use of mechanical harvesting equipment.

- Field must be fairly level without undulations to facilitate smooth operation and uniform stubble length.
- For small reapers and binders, plants must be grown in rows.
- Field efficiency of harvesting machines is high in large fields.
- Water control in rice field is essential to ensure that the fields are drained and are relatively dry at harvest time.

Harvesting Method of Crops

Harvesting of crops like paddy and soybean has to be done carefully as the matured grains easily detach from the earheads/pods and, therefore, cannot be harvested by fast working tools or machines. Bengalgram, green gram, lentil are to be harvested at ground level. Oilseed crops pose different type of problems to engineers for mechanisation of their harvesting. Safflower is a spiny crop and difficult to harvest even manually. In case of sunflower, harvesting is simpler as only flower heads are to be collected. In sesamum crops, pods containing seeds are attached to the main stem and they are mostly raised by broadcasting. This also needs gentle handling. Farmers follow different methods for harvesting of rapeseed/mustard and pigeon pea. Mostly, farmers harvest these crops at branch level, but small farmers harvest these crops at ground level. Harvesting of root crops involves digging, shaking to remove adhering soil, windrowing or stacking and picking. A good root crop harvester should give maximum recovery and cause minimum damage to pods or tubers.

The harvesting methods followed by farmers for major crops are as follows:

Cereal Crops

Wheat and Rice: Harvesting of these crop is traditionally done by using local sickle. Improved serrated blade sickles are also in use. The following machines are available for efficient harvesting of these crops.

— Self propelled walking type reaper
— Reaper binders
— Tractor front and rear mounted reapers
— Combine harvesters

Sorghum: Harvesting by local sickle is the traditional practice followed by farmers. Suitable machines are not available for harvesting this crop. However, combine harvesters are in use in advanced countries.

Maize: The traditional practice is to collect the matured cobs manually. Grain combines equipped with com-head snapping unit are being used in developed countries. Self propelled corn harvestors of different designs are available in developed countries.

Pulse Crops

Bengal gram: Harvesting by local sickle is the traditional practice. Improved serrated blade sickles are also in use. The performance of narrow pitch cutter bar with horizontal conveyor is better than other types of available reapers. Combines with floating cutter-bar are in use in advanced countries.

Pigeon pea: This crop is traditionally harvested at ground level by using a chopper or local sickle. No suitable machine for harvesting this crop is available in the country. Crop stems are being used by farmers for domestic use.

Urad, Moong and Cowpea: These crops are traditionally harvested by using local sickle. Improved serrated blade sickles are also in use.

Oilseed Crops

Groundnut: Digging of crop with country plough and blade hoe at proper soil moisture level and manual pulling and gathering of pods using hand hoe is common practice. Animal drawn and tractor operated diggers and digger windrowers are improved implements developed for groundnut harvesting. The blade harrow is widely used for digging of groundnut crop in Gujarat. TNAU, CIAE and CTAE, designs are some of the improved animal drawn groundnut diggers. Tractor operated groundnut diggers have wide blade, which cover 1.25 to 2 metre width and operate at 10 to 15 cm depth. PAU, APAU and GAU diggers are some of the designs developed.

Rapeseed and Mustard: The traditional practice is to harvest manually using sickles. In tall varieties, farmers cut the plants above-ground level and leave long stubbles in field, which are subsequently ploughed in. In some areas, where plants are used as fuel or thatch material, harvesting with serrated blade sickles close to ground level, is practised by farmers.

Mechanical harvesting of crop is carried out by following type of machines:

— Self propelled walk behind type vertical conveyor reaper,

— Tractor front or rear mounted pto operated reaper, and

— Combine harvester

Combine harvesters need slight modification to make them suitable for harvesting rapeseed crop because of tall and branchy nature of the crop. The changes required are (a) attachment of vertical cutter bar at outer ends to get a clean cut windrow and (b) increase of header table length to accommodate the bulky crop. Combine harvesters can handle the crop easily after above changes are made.

Soybean: Harvesting by local sickle is the traditional practice followed by farmers. However, modified serrated blade sickles are recommended, as plant stem is 8 to 12 mm thick.

The following machines have been evaluated and found suitable for soybean harvesting:

— Self propelled vertical conveyor reaper windrower

— Tractor rear mounted reaper

— Combine harvester

When the available harvesters are to be used for soybean, these are required to be modified and adjusted to reduce field losses and suit crop and soil conditions. Cutting of crop close to ground with low stubble height and crowding and stripping effect are the main requirements. Combine harvesters with floating cutter bars are recommended for low harvesting losses. Narrow pitch cutter bar has been reported to give lower harvesting losses as compared to conventional cutter bar.

Sunflower and Castor: The traditional practice is to manually harvest the flower heads of sunflower and castor plants. These are stacked and sun dried for threshing. Suitable machines are not available for harvesting of sunflower and castor crops. Harvesting of whole plant would require separation of flower heads for threshing and thus the time saved by harvesting the whole plants would not reduce the labour requirement. The combine harvesters are used for harvesting of above crops using specially designed header. These are in use in advanced countries.

Safflower: The traditional practice is to, manually harvest the crop using sickles. Because of thorny and spiny nature of crop, harvesting and handling of safflower plants is a problem. Use of hand gloves and covers on legs and arms is recommended during harvesting. Hayforks are used for gathering and stacking the plants in field or on trailers.

For mechanical harvesting of safflower following machines are recommended:

- Self propelled (1 metre wide) vertical conveyor reaper,
- Combine harvester

Sesamum and Linseed: Traditional practice is to harvest and gather the crop manually using sickles. Tractor rear and side mounted reaper can be used for harvesting the broadcast crop. Vertical conveyor reapers have been used for harvesting crop, raised in rows and at optimum moisture level, i.e. 15-20 per cent, to avoid shattering of pods.

Factors Affecting Performance of Harvesting Machines

Crop Factors

- Crop variety
- Ambient temperature
- Maturity of crop
- Crop moisture
- Crop condition
- Crop density

Machine factors

- Shape and size of crop divider
- Reel position and speed
- Cutting blade shape and speed
- Conveyor speed
- Machine vibrations
- Machine settings

Operational factors

— Height of cut

— Operation speed

Types and Importance of Sickles

Sickle is one of the most common hand tools used for harvesting of the crops, grass and cutting of other vegetative matters. It consists of curved steel edged blade fixed to the wooden handle.

The cutting edge may be plain or serrated depending upon the design. The curvature of the blade differs widely in plain edged and serrated sickles. The blade is fabricated from medium carbon or high carbon steel and hardened to resist wear. Manganese steel, tool steel and alloy steel can also be used to produce better quality sickles which have more life than the sickles fabricated from the plain carbon steel.

The blade has a knife section and the thickness reduces towards the cutting edge. The tang of the blade is inserted in to the wooden handle for joining the blade to the handle. The wooden handle is either straight or bent at the end. For cutting, the part of the plant to be cut is held in one hand and sickle is operated with other hand. Cutting is achieved by imparting translatory and rotary movement to the blade around the point of cut. This action leads to applying frictional and shearing force components required to achieve cutting.

Fruit Harvester

It consists of one fixed blade and another moving blade actuated by a spring. There is a net basket attached to it to collect plucked mango fruits. The long handle facilitates reaching fruits from ground. During harvesting, the fruit pedicel is adjusted to rest on fixed blade and pressing the lever at the grip end of the handle actuates the moving blade. The overall length of the tool is about 3000 mm and the weight of the cutting head is 1.3 kg. The blades are made from carbon steel. The device is also used for harvesting oranges apple and sapota.

Sapota Harvester

It consists of main body of PVC having cylindrical shape. The upper end of the body is closed while bottom end is open to which nylon net for collecting the fruits is tied. A stretched string closes the other end of the net. A gate is made on the body for entry of the fruits to be harvested. On the lower surface of the body a metal holder is fixed to hold the bamboo of required length. Two fingers cut in V -shape and with small sharp blades are provided at the closed end of the body of the harvester. The fingers help to select and hold the fruit to be harvested from the bunch. By pulling the harvester, fruit is detached from the bunch, which falls in the body and rolls into the net. To unload the harvested fruits in the net a stretched string at the closed end of the net is loosened. The harvester is used for harvesting of small fruits like lemon, sapota etc.

Bhindiplucker

The tool is ergonomically designed. The plucker consists of two arms hinged together, cutting blades joined to open ends of arms and two rings joined to the arms. The blades are made of medium carbon steel or low alloy steel, hardened and tempered to suitable hardness. Panicles are cut individually using this tool. The operator is spared of drudgery, discomfort and itching to skin of his hands, which are associated with conventional method of manual plucking without any aid. It fits in to the hand properly with the help of two rings, one over thumb and another over index finger.

Force to cut the pedicle is exerted by pressing these two fingers against each other. Pedicle is sheared between two straight blades, one of which is notched for better grip. It is used for plucking of bhindi (ladies finger) from plant.

Cotton Stalk Puller

It is a simple and useful hand tool for pulling the cotton stalks from the field after harvesting. The tool consists of inverted 'V' frame, footrest, steel jaws and handle. All the parts are made of mild steel and assembled by using fasteners. The movement of the jaws is linked with the motion of handle, which is moved towards and away from the operator. For operation, the stalk in held between the jaws, a foot is placed on the footrest and the handle is pulled. With the pulling of the handle, the jaws grip the stalk and also move it upward and uproot the stalk. The rate of uprooting depends on the moisture content of the soil and skill of the operator.

Lac Sheller

The hand operated lac sheller is provided with two discs mounted on axles. One disc is kept stationary and the other disc having pegs fixed on the inner face can be rotated. It is suitable for removing the resin from all types of lac sticks. It saves 29 per cent labour and 37 per cent on cost of operation and is 2.5 times faster, compared to conventional method of manual scrapping with the help of special knife (dabia). The quality of lac is not affected by the use of lac shellers.

Leafy Vegetable Harvester

It is a modification of hedge shear to which a gathering mechanism provided. The mesh is made of mild steel. The blades of the shear are made from high carbon steel, alloy steel or tool steel. The edges are hardened and tempered to suitable hardness. The blades of the shear are joined to the wooden handle by tang. The handles are shaped for a comfortable grip. The harvester is operated by closing and opening of the blades with both hands. In the open position the stems of the vegetable are placed between the blades—and during closing the vegetable is cut. The cutting takes place due to shearing action. The cut crop is collected in the mesh welded to the blades. The collected vegetable is placed over a twine for binding. One person operates the harvester. For harvesting of leafy vegetables such as spinach and fenugreek.

Rubber Tapping and Draining Knife

The tool consists of a V-shaped, front edged blade, fixed to a wooden handle with the help of tang and ferrule. The blade is made from high carbon steel, tool steel or alloy steel. The cutting edge is heat treated and hardened to—suitable hardness. The ferrule is made of mild steel and chrome plated to avoid oxidation. For tapping of the rubber fluid from the rubber trees, the V -shaped front edge, which is a cutting edge, is used to inflict a cut! groove on the tree trunk through which the rubber fluid slowly flows out. The groove acts like a channel and opens the latex cells and thus allows tapping and draining of the latex. A tin is placed at the end of groove or channel for the collection of the tapped fluid! latex. Fonnation of the groove is due to the scratching and shearing action of the tool head. The blade is made from carbon steel and is heat-treated to obtain a hardness of 30-45 HRC. This tool is used for tapping fluid from rubber trees.

Animal Drawn Potato Digger

It is an animal drawn digger useful for digging and exposing potato tubers. It consists of a ridger shaped bottom with welded extension rods on its wings. These rods help in separation of soil and dirt from the potato tubers. A handle is provided at the rear for guiding the implement while in operation. It is suitable for digging potato tubers after removal of vines from the field. It saves 40 per cent labour and operating time and 18 per cent on cost of operation compared to conventional method of digging with spade. It also results in reduction of 11.3 per cent losses compared to conventional method of digging with spade.

Threshing Equipment

Thresher is a machine to separate grains from the harvested crop and provide clean grain without much loss and damage. During threshing, grain loss in terms of broken grain, unthreshed grain, blown grain, spilled grain etc. should be minimum. Bureau of Indian Standards has specified that the total grain loss should not be more than 5 %, in which broken grain should be less than 2 %. Clean unbruised grain fetch good price in the market as well as it has long storage life.

Trampling of paddy under feet, beating shelves of rice or wheat crop on hard slant surface, beating crop with a flail, treading a layer of 15 to 20 cm thick harvested crop by a team of animals are traditional methods followed by farmers depending upon capacity, lot size and situation. Threshing by bullock treading is practised on large scale in the country but it is also time consuming and involves drudgery. Tractor in many places is now used in place of animals for treading. Introduction of animal drawn olpad thresher reduced the drudgery of the operator and gave comparatively higher output per unit time. In all above methods the threshed materials are subjected to winnowing either in natural wind flow or blast from winnowing fan for separation of grain from straw.

Threshing wheat by traditional method involves drudgery and takes more time to obtain required quality of bhusa. Due to these, mechanical threshers are widely accepted by the farmers.

A mechanical thresher consists of the following parts:

- Methods Feeding device (chute/tray/trough/hopper/conveyor)
- Threshing cylinder (hammers/spikes/rasp-bars/wire-loops/syndicator)

- — Concave (woven-wire mesh/punched sheet/welded square bars)
- — Blower/aspirator
- — Sieve-shaker/straw-walker.

The crop is fed from the feeding tray into the threshing cylinder. The threshing cylinder is fitted with spikes/bars/hammers or wire-loops around its periphery according to the type of thresher. Below the cylinder there is a concave and it covers lower portion of the cylinder. The cylinder rotates at high speed and thus the crop is threshed and the entire or a portion of threshed material falls from the concave on' to top sieve of cleaning system. Due to reciprocating motion of top sieve lighter material accumulate at the top and grain falls on to the bottom sieve. In case of spike-tooth thresher, an aspirator blower sucks out the lighter material from the top sieve and throws it out from blower outlet. The sieves help in further cleaning of the grain by allowing heavier straw to overflow.

Various adjustments are required before starting threshing operation. The machine is to be installed on clean level ground and is to be set according to crop and crop conditions. The adjustments necessary to get best performance from the machine are

- — concave clearance,
- — sieve clearance,
- — sieve slope,
- — stroke length and
- — blower suction opening.

Besides these, cylinder concave grate, top sieve hole size and cylinder speeds for threshing different crops are important for a multicrop thresher.

Principle of Threshing

During operation, the crop material is slightly pushed into the threshing cylinder through the feeding chute, which gets into the working slit created between the circumference of revolving drum having attached spikes and the upper casing. The speed of the spikes is greater than the plant mass due to which they strike the latter which results in part of the grain being separated from straw. Simultaneously, the drum pulls the mass through the gap between the spikes and the upper casing with a varying speed.

The angle iron ribs on the other hand, restrain the speed of the travelling of stalks clamped by the spikes. Due to this the spikes move in the working slit with a varying speed in relation to the shifting mass of material, which is simultaneously shifted, with a varying speed with respect to the upper casing. As a result, the material layer is struck several times by the spikes against the ribs, causing threshing of the major amount of grains and breaking stalks into pieces, and also accelerating them into the inlet of the lower concave.

As the material layer shifts towards the progressively converging slit of lower concave, its size reduces. The vibration amplitudes therefore, decrease, whereas the speed of the layer increases. This causes mutual rubbing of the ear stalks, as well as rubbing of the ears against the edges of the concave bars and causes breaking of stalks depending on the concave clearance. Since the system is closed, the thicker stalk, which cannot be sieved through the concave, again joins the fresh stalk and the same process is repeated until the stalk size is reduced to the extent that it can pass through the concave apertures. Thus fine bruised straw is produced.

The effective threshing process means that the loss of unthreshed kernels ejected with the straw through the concave and the loss of grain damage should be low and the amount of the material passed through the concave should be high.

Type of Thresher

The type of thresher is generally designated according to the type of threshing cylinder fitted with the machine. The major types of threshers commercially available are as follows:

- Drummy type: It consists of beaters mounted on a shaft which rotates inside a closed casing and concave.
- Hammer mill type: It is similar to dummy type but it is provided with aspirator type blower and sieve shaker assembly for cleaning grains.
- Spike-tooth type: Spikes are mounted on the periphery of a cylinder that rotates inside a closed casing and concave. It is provided with cleaning sieves and aspirator type blower.
- Raspbar type: Corrugated bars are mounted axially on the periphery of the cylinder. It is fitted with an upper casing and an open type concave at the bottom of the cylinder. The cleaning system is provided with blower fan and straw walker.
- Wire-loop type: Wire-loops are fitted on the periphery of a closed type cylinder and woven wire mesh type concave is provided at the bottom.
- Axial flow type: It consists of spike tooth cylinder, woven-wire mesh concave and upper casing provided with helical louvers.
- Syndicator type: The cylinder consists of a flywheel with corrugation on its periphery and sides, which rotates inside a closed casing and concave. The rims of the flywheel are fitted with chopping blades. In case of wheat threshing farmers want not only clean grain but also need fine quality of bruised straw (bhusa) for cattle feed. This requirement of quality 'bhusa' makes combining unattractive to small farmers.

Drurnmy type, hammer mill type and syndicator type threshers are suitable for threshing wheat crops only and they can produce fine quality of 'bhusa'. Rasp-bar type, wire-loop type and axial flow type threshers are suitable for paddy and they do not make fine straw. Rasp-bar type threshers can be used for threshing other crops but farmers do not prefer this machine because it does not make fine 'bhusa'; and cost is very high due to its bulky size. Though the hammer mill type threshers can produce fine quality 'bhusa' its use is decreasing day by day due to high power requirement.

Portable wire loop type paddle operated threshers are widely used by farmers in paddy growing areas. Spike tooth type thresher can thresh wheat crop and can produce fine quality of 'bhusa'. This thresher can be used for threshing other crops if the blower is mounted on a separate shaft so that the cylinder speed can be varied independently. Majority of farmers prefer spike tooth type threshers because of their simplicity in design, low cost and their ability to make fine 'bhusa'.

Factors Affecting Thresher Performance

The factors which affect the quality and efficiency of threshing are broadly classified into following three groups:

- Crop factors:
  - Variety of crop,
  - Moisture in crop material.
- Machine factors:
  - Feeding chute angle,
  - Cylinder type,
  - Cylinder diameter,
  - Spike shape, size, number
  - Concave size, shape and clearance
- Operational factors:
  - Cylinder speed,
  - Feed rate, method of feeding,
  - Machine adjustments.

Beans are more susceptible to damage due to impact and the variety of grain has much influence on grain loss during threshing. Damage of large beans is more than smaller beans at same impact velocity and orientation. The amount of damage increases rapidly below ambient temperature of 10°C. So, handling of dry beans at low temperature should be avoided. Moisture content of grains is a major factor in controlling grain damage.

Decrease in moisture content greatly increases the brittleness of grains. Unthreshed grains are more at high pod moisture content whereas grain damage decreases with increase in grain moisture content. More threshing effort is required for threshing high moisture crop, which causes more internal grain damage and thus affects viability. For Soybean moisture content between 8 and 12 % (wb) is optimum for low mechanical damage.

The base angle of feeding chute affects the feed rate. It should be tangential to cylinder drum for maximum feed rate and minimum physical effort. The threshing cylinder requires power as high as 60-75 % of total power input. Hammer mill type threshers bruise the straw very fine but the specific energy requirement is highest among all types of threshers. Rasp-bar cylinder design can thresh most of the crop except groundnut but these machines do not provide bruised straw.

The concept of a straw bruising attachment to rasp-bar thresher is not economically viable. Spike tooth type threshers having independent drive to cylinder and blower can thresh major crops effectively but the cylinder speed is to be adjusted according to the crop conditions. Larger cylinder diameter has lower power requirements than smaller ones at higher feed rates. Higher rib spacing in upper concave increases unthreshed grain but reduces power consumption. The performance with flat spikes is better than round and square spikes. Larger spike spacing in a row reduces power consumption and broken grains whereas power increases and broken grains reduce with the increase in number of rows of spikes. However, uniformity of spike distribution over cylinder periphery is more important for better performance.

Power consumption and grain damage increases with the increase in spike length and thickness. The grain damage decreases and unthreshed grains increase with the increase in concave gap. Higher concave clearance reduces power consumption whereas straw bruising is more at low concave clearance.

The effect of cylinder speed on threshing performance is highly significant at all machine settings. Power consumption and broken grains increase and unthreshed grains decrease with the increase in cylinder speed. Though the unthreshed grain losses decrease but the total grain losses increase with the increase in cylinder speed. Quality of bhusa is better at higher cylinder speeds, low concave clearance and concave gap. Higher feed rate increases power consumption, reduces broken grains and unthreshed losses and to some extent helps in straw bruising. In general, feed input capacity is considerably affected by machine settings. Lower rib spacing in upper concave, concave bar spacing, concave clearance and non-uniform spike distribution over cylinder periphery reduces the capacity. Low feed rate, high labour rate, high energy consumption, high percentage of broken grain and poor quality of straw all contribute to high qualitative cost of operation.

Pedal Operated Thresher

It consists of wire-loop type threshing cylinder, power transmission system, mild steel sheet body and foot pedal. The threshing cylinder consists of wire-loops of I '0' shape embedded in wooden or metallic strips joined to two discs. A shaft carries the threshing cylinder and is connected to the transmission system. The transmission system consists of meshed gears or sprocket-chain mechanism. The larger gear or sprocket is connected to foot pedal/bar with links. The foot pedal/bar is always in raised position. On pressing the pedal the threshing cylinder starts rotating. For continuous rotation of the cylinder, the pedal is lowered and raised repeatedly. For operation, paddy bundle is held in hands and earhead portion of the crop is placed on the rotating cylinder. The wire-loops hit the earheads and grain get detached from the rest of the crop. It is used for threshing rice crop.

References

Chatha, I.S. and Grewal, S.S., *A study of tractorisation in Punjab*, Department of Economics and Sociology, Punjab Agricultural University, Ludhiana, 1991.

Datt, S. and Sundharam, K.P.M., *Indian Economy* S. Chand & Co. Ltd., 1998.

Mishra, P. and Sundram, *Some aspects of the economics of harvest combines in Punjab*, Economic & Political Weekly, 10(39), 1975.

12

Temporary Fencing Techniques

Improved solid state technology and developments in wire for fences have created new possibilities for livestock and pasture management using rotational grazing. A good understanding of the fencing materials that are currently on the market will help in selecting, installing and maintaining the rotational grazing system.

Temporary Electric Fence Wires

There are two basic types of temporary electric fence wires in the market: polywire and polytape. Polywire is a generic term referring to any of several brands of electroplastic twine. Polytape refers to electroplastic ribbon. Polywires and polytapes are made up fine metal filaments braided with strands of polyethylene or polypropylene fibers. The metal carries the shock and the plastic gives the temporary wire strength and visibility.

Most producers subdivide their pastures with polywire instead of polytape. The primary reason is cost. Polywire costs about half as much as polytape. It typically has three, six or nine strands of fine metal filaments that carry the shock. Some manufacturers use aluminum filaments, which make a better electrical conductor compared to typical polywires with stainless steel.

The disadvantage of using aluminum is that it can "work harden" over time, which makes the metal brittle and subject to breaking. If it is planned to put the fence up and leave it in place, then polywire that uses aluminum filaments will do an excellent job of controlling livestock. However, if it is planned to move the fence to adjust the paddock size, which is often the case with rotational grazing, polywire with stainless steel filaments will be more durable and last longer.

Stainless steel doesn't work harden like aluminum, but it is a slightly poorer electrical conductor. As a result, most basic polywire has six conductive filaments. Some manufacturers also sell polywire with nine strands. It costs more than six-strand polywire, but it has 50 percent more conductivity.

Some manufacturers sell nine-strand polywire that contains both stainless steel and highly conductive alloy filaments. It costs more per meter compared to six-strand polywire, but is 100 times more conductive. If polywire is used for short distances to subdivide a pasture, six-strand polywire is sufficient to do the job.

Polytape typically consists of anywhere from 5 to 15 strands of metal filament woven with strands of plastic to form a flat ribbon. Most polytapes on the market use stainless steel filaments. Polytapes are more visible than polywires.

If animal control is poor with polywire or visibility is important, such as with horses, it might be consider switch to polytape. Having better animal control will offset the additional cost of polytape. Polywire and polytape won't last forever.

If an animal doesn't see the wire, it can't respect it. Visibility is critical with temporary fences, especially with poorly trained stock or where wildlife may challenge the fence. Polywire comes in several colors and color combinations: white, orange, orange and black, yellow, and yellow and black. Against a lush green or dried yellow background or at night, white is more visible. Time and time again untrained animals notice white from farther away than any other colors.

Kinds of Posts

There are four basic kinds of posts generally used for temporary fences. Plastic posts are more convenient to use where the fence is to be moved frequently, because they have built-in treads to "step" the posts into the ground. The pre-molded loops provide plenty of flexibility for a variety of wire spacings. They are slightly more expensive than other temporary posts and have a life expectancy of three to five years.

Fiberglass rods are better suited for situations where you won't be moving the fence very often. They are less expensive than plastic posts and can last up to 20 years. They are easily driven in the ground with a hammer, but they do take a little longer to install than plastic posts. Fence suppliers sell a cap to place over the end of the rods to keep them from splintering when hammered.

A spent shotgun shell also works well and doesn't cost anything. Some producers have made a mini-post driver from a ½-inch diameter piece of steel pipe with a cap on one end. Wire clips or plastic insulators that slide on the rods hold the wire in place.

Many producers use 4-foot long, 3/8" metal rebar posts. They are cheaper and last longer than plastic or fiberglass posts. They are driven in the ground with a hammer. Slide on plastic insulators hold the wire in place. They are not flexible like plastic or fiberglass posts and will bend if the fence is hit by animal.

Metal "t" posts are stronger and last longer than the other temporary posts, but they cost more and require more labor to install and remove when changing the paddock size. However, since they are not flexible like the other posts, they do work well as corner posts or when making sharp turns in the fence line.

Types of Electric Fence Chargers

Touching an electric fence is very unpleasant. The experience for animals is the same. The severity, or the amount of shock the animal feels, depends on the voltage and amperage as well as the duration of the shock. It takes a minimum of 700 volts to produce a shock sufficient enough to control short-haired breeds of cattle, pigs and horses, and a minimum of 2000 volts for long-haired cattle, sheep and goats.

Commonly, the best fence chargers will deliver 4500 to 7000 volts. It is important to realise that voltage level only determines whether or not a shock will be delivered to the animal, and is not a measure of how much shock will be felt. If the voltage is adequate to deliver a shock, the effectiveness will then depend on how much shock is produced (shock intensity).

The energy output in a shock pulse is an indicator of shock intensity. It is related to the combination of amps, volts and pulse on-time, and is measured in joules. A joule is a measure of energy equal to a watt for one second. High joule values indicate high shock intensities. The joule rating can range from less than 1 for small battery-operated chargers to about 40 for the most powerful models. Remember, double the joules and you'll double the shock. There are two types of electric fence chargers on the market: high- and low-impedance. Impedance means leakage.

Generally speaking, high-voltage chargers with long pulses are high- impedance chargers. Chargers with long pulse lengths can cause current to arc and heat to build up in the wire. This can cause fire and melt polywire and polytape where it comes in contact with weeds and grass. Avoid using high-impedance "weed-chopper" type chargers. These chargers leak current too readily.

Any weeds or grass that touch the fence will drain the power to the extent that little or no shock is felt by the animal. Low-impedance chargers are better suited for use with rotational grazing systems because they resist leakage. Because they resist leakage, they have the capacity to power long distances of single or multi-wire fence. Low-impedance chargers put out an intense pulse which lasts for 0.0003 seconds or less, which eliminates heat buildup in the wires.

If fields are near a 120-volt power source, the best choice is an AC-powered charger. AC chargers usually have higher joule ratings and require no battery maintenance. Operating costs are reasonable, averaging Rs. 30 to 40 a month.

If the fence being planned to electrify is not near a 120-volt power source, there is no choice but to use a battery-operated (DC) charger. Low-impedance DC chargers do an excellent job of confining animals and are very dependable. Some DC chargers have a battery and recharging system (photovoltaic panel) built into the unit. Others operate on a separate 12, 24 or 36 volt system. With these systems you can either recharge the batteries with a standard battery charger every two to six weeks depending on the type of charger and the amount of use, or you can buy a solar pack to keep the batteries charged. Deep-cycle batteries are best suited for battery-operated chargers. They can be discharged and recharged repeatedly. Batteries designed for use in automobiles are not designed to be discharged and recharged and thus will not last as long as deep-cycle batteries. Once totally discharged, car batteries will only recharge up to 60 to 75% of their original capacity.

Grazing System Layout

Water

Water is an essential part of a grazing plan. The distance to water is an important consideration when designing a rotational grazing system. The shorter the distance animals have to travel to water, the more water they will drink and the less energy they will burn traveling back and forth from the paddocks to the water source. Both of these factors will have a positive influence on animal performance. Ponds, wells, creeks, springs or public water sources can be used to supply water to livestock. Regardless of the source, it must be capable of supplying enough clean water throughout the entire grazing period to meet livestock needs. The method by which water is supplied to animals in a rotational grazing system will vary greatly from farm to farm and may even vary on an individual farm.

Two systems that can be used to bring water to livestock are permanently installed water lines running to a series of water troughs in the paddocks, or temporary water systems, such as plastic pipe on top of the ground that uses portable tanks. If it is not feasible to bring water to the livestock, then fencing can be designed and installed to move animals from the paddocks to a water source. The most common method is to build a lane that joins a series of paddocks to the water source.

Structure of Paddocks

Livestock like to travel fence lines to locate their boundaries or places where they might escape. In doing so, available forage is wasted by trampling and deposition of manure and urine. Square paddocks (or shapes that approximate a square) are the most efficient compared to other shapes (pie-shaped, fan-shaped, narrow rectangles) because animals can obtain their daily feed with a minimum of time, effort and trampling of the pasture. Rectangular paddocks may also be used as long as they are no more than four times as long as they are wide.

If the paddock is long and narrow, cattle will graze closest to water and shade then graze young regrowth from the same area before intensively grazing the far end of the paddock. Although sometimes it is necessary to use other paddock configurations, in particular when fence lines have to follow natural land forms or boundaries, the use of circles, triangles or other shapes should be kept to a minimum.

Avoid paddocks that funnel down to a narrow area at the gate. This concentrates animals to a narrow lane arrangement and increases the tracking and erosion problems. Wherever possible, fence across rather than up and down slopes.

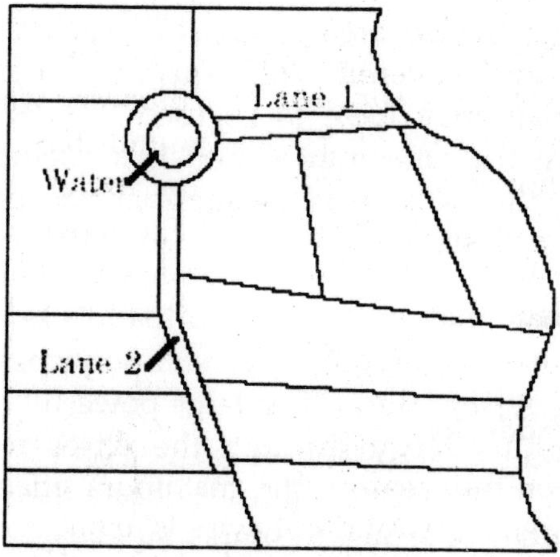

Figure 1. Paddock Layout

Gates

Locate gates so they do not interfere with the natural movement of livestock as they travel to and from the barn, water, shade or handling facilities. Generally, gates should be located in the corner of the paddock that is closest to the direction the livestock need to travel. If not, although some of the livestock will find their way out of the paddock, some animals will end up trapped in a corner without a gate trying to figure out how to tear the fence down.

Lanes

Lanes should be constructed so that livestock can be easily moved from paddock to paddock, to the water supply and to the handling facility or barn. Locate lanes on a contour where little change in elevation is necessary along the route. This will help prevent gullies from forming along the animal traffic lanes. It is important to make lanes wide enough to spread the animal traffic out and to get hay-making or other machinery through, but not so wide that valuable grazing land is wasted. A width of 16 to 24 feet for lanes is usually sufficient.

In heavy traveled areas, geotextile material anchored to the ground and covered with crushed stone, concrete or other materials may have to be used to prevent livestock from turning the lane into a mud hole. Being able to get livestock to the pasture is as important as producing the forage in the pasture.

Installation

The fine wires used in polywires and polytapes have a high resistance to current flow. Even large powerful chargers can't pump enough energy through the wires for very long distances. For this reason, the maximum effective distance you can charge a 6-strand polywire is about ½ mile from the charger. You can power up to a mile of polywire by placing the charger in the middle of the fence and going ½ mile in each direction. It is best to run a single, electrified-strand of barbed wire or high-tensile wire along the perimeter of the pasture and run the temporary wire off this feeder wire when subdividing large pastures. Do not use standard polywire or polytape as a feeder wire because it won't carry a sufficient charge. Never put two chargers on the same fence.

Generally, only one strand of polywire is needed for cattle and horses (sheep may need two or three strands). The wire should be installed 30-32 inches off the ground. This spacing may not hold calves effectively; however, it won't upset the cows as long as the calves are within sight. It may actually prove to be a benefit since calves can graze on the fresh forage in front of the cows. Be careful not to overstretch the wire. If it is overstretched, the metal filaments will break and short the wire. Depending on the terrain, posts can be up to 100 feet apart.

Grounding Electric Fence System

A majority of electric fence problems are the result of poor grounding. For an animal to receive a shock, it must complete a circuit. With temporary electric fences, this circuit consists of one "hot" wire and a soil ground return. The charger condenses electricity to generate a high-voltage pulse. From there, it travels through the wire, through the animal, to the soil back to a ground rod and then up the ground rod back to the charger. That is when the animal feels the shock. The grounding system for an electric fence is a little like a radio antenna. With a radio, the bigger the antenna, the better the reception.

Likewise, your electric fence charger requires a good grounding system to collect enough electrons from the soil to complete a powerful circuit and produce the shock felt by the animal. The shock felt by the animal from a poorly grounded system in most cases will be inadequate to keep the animal from walking through the fence.

The first step in establishing a good grounding system is to check the manufacturer's recommendations. Most manufacturer's recommend a minimum of three, 6-foot long galvanised ground rods to properly ground their electric fence chargers. More powerful units may require up to eight or 10 rods. They should be driven 5 ½ feet into the soil and spaced a minimum of 10 feet apart (Figure 2). Locate ground rods so they are:

— highly visible,
— near permanent moisture,
— not likely to be damaged from or cause injury to livestock,
— do not interfere with mowing or cultivation.

Where rocks prevent driving rods, bury a long (20+ feet) galvanised pipe in a trench as deep as possible for a grounding sytem. Connect the charger to the ground rods with an insulated cable or 12 ½-gauge galvanised wire. Use one continuous wire and firmly attach the wire to each rod with ground rod clamps. Simply wrapping a wire around the ground rod makes a poor connection. Do not install your grounding system within 50 feet of any utility company

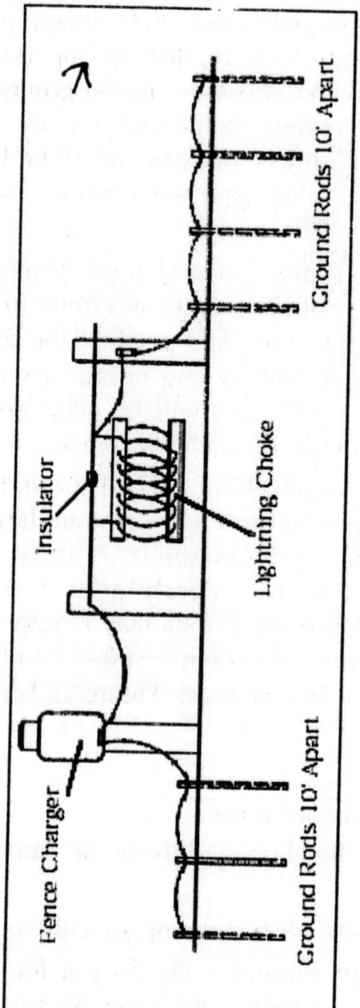

Figure 2. Grounding Procedures for Electric Fence Chargers

ground rods, telephone ground rods or any underground telephone and power lines to reduce the chance of lightning damaging your charger.

Testing the Effectiveness of Grounding System

Adequacy of current grounding system can be tested using a voltmeter and the following procedure:

- Lean several metal "t"posts against a hot fence wire and the soil about 100 yards from the ground rods to simulate heavy vegetative growth. Use enough metal post so that your voltmeter shows less than 1,000 volts in the hot wire.
- Connect one lead from voltmeter onto the ground rod furthest from the charger.
- Make a firm connection with the other voltmeter lead to the soil (it may help to push a stiff wire several inches into the ground and connect the voltmeter lead to the end of this wire).
- If the voltmeter reads more than 500 volts, then there may have the following problems:
    - Not enough ground rods (this is the most com mon problem),
    - Ground rods too close together,
    - Poor connections between ground rods.

References

Chatha, I.S. and Grewal, S.S., *A study of tractorisation in Punjab*, Department of Economics and Sociology, Punjab Agricultural University, Ludhiana, 1991.

Datt, S. and Sundharam, K.P.M., *Indian Economy* S. Chand & Co. Ltd., 1998.

Gajja, B.L.; Kalla, J.C. and D.L. Vyas, *Determinants of tractorisation in arid areas of Western Rajasthan*, Agricultural Situation in India, 40(4):257-260, 1985.

Kumar, N. et al., *Farm mechanization and its impact on Punjab agriculture*, paper presented in I.S.A.E. Seminar report, Khalsa College, Amritsar, 1995.

Lekhi, R.K. and Singh, Joginder, *Agricultural Economics*,Kalyani Publishers, New Delhi, July 2002.

Mishra, P. and Sundram, *Some aspects of the economics of harvest combines in Punjab*, Economic & Political Weekly, 10(39), 1975.

PAU, Ludhiana, Directorate of Extension Education, *Package of Practices for Rabi Crops* in Punjab, 2002.

Pathak, A.S., *Farm mechanization and productivity*, Productivity, 23(4): 411-417, 1985.

Planning Commission, Govt. of India, *Expert committee on combine harvesters*, Report (Mimeographed), 1975.

Sethi, K., *Optimum land use pattern and normative demand for tractors in a growing economy-a case study of Ludhiana district*, unpublished M.Sc. Thesis, Punjab Agricultural University, Ludhiana, 1992.

Sharma, B.M.; Goyal, S.K. and G.P. Sharma, "Input use efficiency on tractor and bullock operated farms in Aligarh district of Uttar Pradesh", *Indian Journal of Agricultural Economics*, 47(3): 487, 1992.

Singh, B., *Demand for tractors in Punjab*, unpublished M.Sc. thesis, Punjab Agricultural University, Ludhiana, 1979.

13

Future Requirements of Agricultural Mechanisation

At the time of independence in 1947, Indian agriculture used mostly bullock drawn ploughs, wooden pegtooth harrows, wooden planks for pulverisation compaction and smoothening, bullockcarts, and handtools such as Khurpi, crobar, spade, hoe, sickle, axe, chopper/Dau etc. Some philonthrops, facilitated by colonial masters, were trying to introduce western soil inversion plough like Meston plough, a 15 cm size long beam bullock drawn plough. The only mass manufactured items were spades, pickaxe, crobars, and watering buckets manufactured by Tatas. Traditional equipments in use were made by local craftsmen, using locally available materials with little standardisation, where quality depends on the craftsmanship.

Agricultural Mechanisation in India

Colonial India had witnessed many droughts and famines and some of them very devastating. Founding fathers of modern India took upon themselves a responsibility to avoid famines. To this effect they accorded high priority to agriculture and allied activities. Agricultural education, research and development infrastructure were strengthened. State Agricultural Universities on the pattern of American Land Grand Colleges with integrated teaching, research and frontline extension were established starting 1960 providing trained human resource, location specific technologies and their frontline demonstrations. Indian Council of Agricultural Research (ICAR) reorganised itself. The commodity institutes with the Deptt. of Agriculture and Cooperation and Central Commodity Committees were transferred to ICAR leaving tea, coffee, silk etc.

Central and State Governments invested in infrastructure development to provide irrigation water, seed, fertiliser, pesticides and credits, minimum support price (MSP) and buffer stocking to prevent wild fluctuation in market prices besides food security. However, food security eluded India for two decades, food imports at times being the largest import bill of the country. India witnessed Green Revolution in 1967-68 and subsequently White Revolution, Blue Revolution and Yellow Revolution, achieving quantum jumps in production and productivity in foodgrains, milk, fishery, and oilseeds, respectively. It is experiencing horticultural revolution, revolution in poultry production. Peers and policy makers are looking towards Evergreen Revolution.

Developments in Agricultural Mechanisation

These amendments in agriculture were accompanied with due inputs of mechanisation in natural resource development, agricultural field operations and on-farm primary processing. After intensive testing and evaluation in late 1950s, manufacturing of irrigation pumping sets commenced. Initially two-thirds were engine operated and one-third electric operated. As rural electrification advanced, proportions have changed in favour of electrical power.

Animal drawn improved equipment such as seed drills, seed-cum-fertiliser drills, 5 hp power threshers primovers like diesel engines, electric motors got into manufacture and use. Central Tractor Organisation (CTO) established soon after independence to reclaim marshy lands in Tarai of UP and scrub forests elsewhere to settle displaced people who came from across the border set the pace of tractorisation in India. CTO used crawler tractors, their operation, upkeep and later on indigenous fabrication of certain fast wearing components, after OE stocks exhausted, were locally developed. For tractorisation of agricultural field operations around mid-1960s small 4-wheel general purpose tractors were brought in CKD (completely knocked down) condition and assembled, marketed, operated, and serviced by training Indian technicians.

Confidence thus gained resulted in progressive indigenous manufacture. Swaraj 35 hp from M/s Punjab Tractor was the first totally indigenous tractor. A little later two-wheeled tractors popularly known as power tillers were introduced and at one stage more than a dozen firms had manufacturing licenses. However farming system in vogue, wet cultivation during kharif and upland farming during rabi, and lack of proper after-sales-services support adversely affected their growth. All but two Mitsubishi and Kubota by M/s VST Tillers, Bangalore and M/s Kerala Agro Industries Corporation survived.

Today India is the largest producer of tractors in world with about 2,75,000 tractors per year and about 15,000 power tillers. China is able to market its power tillers in India at cheaper prices, nevertheless there are after sales service problems in many cases. With the introduction and growth of tractors in India in production of matching equipment for scraping and land levelling, seedbed preparation, seeding and planting, seed-cum-fertiliser drilling, spraying and dusting, harvesting and power threshing, 2-wheel and 4-wheel tractor trolleys got in to indigenous manufacture and these got reserved for Small Scale Industries (SSI) sector. It became a very competitive farm equipment industrial activity in Punjab and in pockets all over the country. However, to enhance quality of farm equipment many items have been dereserved, now. By early 1980s vertical conveyor reapers (VCR) were introduced to mechanise sickle harvesting, initially walking type, then a larger tractor version and subsequently riding type self propelled units.

During 1982-84 production of tractor mounted VCRs increased ten fold each subsequent year reaching to 3000 in third year but got reduced to 2000 annual production in the fourth year, the year insurgency in Punjab touched its peak. At this point of time Punjab farmers found combining of rice and wheat cheaper and less risky. Several manufacturers in Small Scale Sector took to general purpose standard grain harvesting combines by manufacturing tractor mounted, self propelled and tractor driven versions. Combining, however, created problem of rice and wheat straw gathering, transforming and handling as Bhusa.

Straw disposal through incineration was found creating serious environment pollution whereas straw incorporation in to the soil was leading to nitrogen stealing. Invention and introduction of straw combines did provide a solution to reclaiming wheat Bhusa but still about 50-60% of the rice and wheat straw is being disposed by burning. It may not be entirely due to combines, demand for wheat bhusa has also declined. Its transport to feed deficit areas in loose farm is expensive and uneconomical. Complete feedblock bufferstocking to fight feed famines is a possibility.

Modern Imperatives

Mechanisation is now demand driven. With increasing labour wages and agricultural produce market prices as they are, engineered to be low for food and nutritional security and food accessibility to the masses, the farmers, specially medium and large ones, are looking for labour saving devices to remain competitive more so with the globalisation of the world markets. And the farm working groups, policy makers and social scientists are looking for mechanisation to remove drudgery from farm operations so that rural educated youth do not run to urban areas in pursuit of jobs which are already in short supply. With prepondence of marginal and small farms, below 2 ha above 76% of the total land holdings, custom servicing in farm operations is in vogue and rightly so because individual ownership is not affordable.

For the reasons explained, agricultural mechanisation in India has come a long way in the last 50 years. India produces wide range of agricultural equipment needed to practice modern intensive farming. However, there are unit operations in certain commercial crops and commodities where mechanisation is needed, and for which presently there is no viable solution. With quantitative restrictions removed from 1st April, 2001 it is faced with new challenges, some have reservations whether it will be able withstand pressures of multinational companies and countries that are aggressively marketing their goods in export markets. Indian farm equipment industry has demonstrated its resilience and responsiveness to changing market situations in the past and should come out victorious at the end adopting modernisation measures.

Nevertheless, the challenge is formidable. To meet food, feed, fibre, fuel, and industrial raw material needs Indian agriculture is required to double food production in a decade i.e. by 2010. With the available land mass remaining at about 142 million hectares (Mha) it has to come through essentially a vertical expansion, gains through productivity in commodities and the regions there is scope like E-India and returns on investment favourable.

Globalisation and its basic demand of competitiveness asks for greater timeliness, precision in metering and placement of inputs that are going to be increasingly costlier, minimisation of pre and post harvest losses, on-farm value addition for additional income and employment that provide greater sustainability to farm families and make farming and associated post-harvest activities less arduous and economically rewarding and satisfying. Environment control in plant and animal production will have to be employed to realise the productivity levels targeted and prevent morbidity and mortality in livestock husbandry.

Status of Farm Power

Unlike industry where men machines and materials are brought under one roof, the agriculture requires men, machinery and material being moved to field and various unit operations performed timely, with required precision, that gives desired productivity in a manner it does not cause drudgery and health hazards to the workers. In this context availability of appropriate farm power sources are required. Traditionally humans and draft animals met this need. Draftability of draft animals has been studied. Bullocks and he buffaloes pull about 10% of their body weight, camels about 18% of the body weight and donkeys about 34%.

With improved yokes and harnesses and appropriate work rest cycle draftability and work output can be increased by 15- 30%. However, for the current level of intensity of farming, required levels of productivity, with a work environment required can not be met by animate sources alone. As a result electro-mechanical sources supplement and substitute animate sources. For stationery operations like operation of irrigation pump, power thresher, chaff cutter and grinder engines and motors are in use.

Where as for tractive field operations there are tractors and power tillers run by diesel engines mounted on them. These can perform varied additional functions with the help of proper attachments. Large farm machines are often self propelled with built in power source. These farm power sources are commercially available but their distribution and availability and use in certain regions like E and NE and hilly areas can not be considered satisfactory. There is close nexus between farm power availability and agric productivity

The level of farm power availability during 2001 was about 1.35 kW/ ha. At the present level of sophistication in agriculture for taking two crops per year an average farm power available of 2.0 KW/ha is considered essential. Agriculturally advanced states and regions in the country have farm power availability of 2-4 kW/ha. Law of diminishing return is also applicable in farm power availability and mechanisation. Excessive mechanisation does not bring rewards instead block the capital, add to overhead costs and at times cause indebtness and its negative impacts.

Farm Implements and Machines

Land and Water Resource Development Equipment: It involves clearing of land from the unwanted vegetation, levelling and grading so as to perform field operations properly and apply seed, fertiliser/manure, and irrigation water uniformly. It also involves creating access to field and mobility of machines through farm roads with irrigation and drainage ditches, creating irrigation water resources both surface and ground water. Most ancient practice of land development was use of fire and later on slash and burn a practice still vogue in a few remote areas. However, axe, hoe, pickaxe, spade, crowbar, and heavy duty ploughs have come in use and still in vogue where mechanisation has not advanced.

Subsequently not long ago animal drawn buck scraper, floats, V-ditchers, bund formers were developed and got in use for levelling, smoothing, land shaping. With industrial revolution and development of IC-engines and advancements in mechanisation the tractor drawn scrapers, levellors, terracers, floats, smoothers have come in use and are available in the market. However, for the large scale mechanised land clearing, land development, construction of farm ponds, laying of irrigation and drainage systems there are bulldosers, motor graders, excavators, dumpers, trenchers, trenchercum-subsurface drainage pipe systems available in the Indian market.

Heavy earthmoving machinery are costly. Laser guided land levellors are in the process of introduction in India and may be indigenised soon. For laying tubewells to tap ground water (GW) there are percussion and rotary rigs along with casing and strainers of steel and plastics and submersible pumps available. Where GW tables are low and rich aquifers available at lower depths centrifugal pumps installed in dugwells are in use. Shallow tubewells in many states rich in GW have paid rich dividends like Assam, West Bengal, Punjab, Haryana, Western UP, and pockets elsewhere.

Seedbed Preparation: Leaving deltaic areas of the perennial rivers where rice-rice may be dominant crop rotation, most of India has crop cultivation in both upland and low land situations. For marginal farms, traditional animal drawn wedge plough which can be used as primary and secondary tillage implement, with a seeding tube as a seed drill and as an interculture device continues to be still in use. With over 76% holdings of marginal and small size (1-2 ha) dominating the scene, it is not surprising that the Desi Hal and other traditional equipment continue to be in use. Wooden plank as an accessory does pulverisation, land smoothening and seedbed compaction.

In vertisols Bakhar are used. In wet land seedbed preparation, wooden pegtooth harrows are also used. For primary tillage, better weed management and incorporation of organic and green manuring, bullock-drawn mould board ploughs of 10-20 cm size, long and short beam types, reversible MB ploughs were introduced and are still in use. For secondary tillage animal drawn disc and pegtooth harrows were introduced and got in to indigenous manufacture. However, these could never gain a popularity as compared to Desi Hal which is a versatile multipurpose implement, made and repaired by local craftsmen. For wet land seedbed preparations there are various animal drawn puddlers which increase field capacity. Agricultural mechanisation process is gradually changing the scenario.

In Punjab, Haryana traditional farm equipments have practically disappeared. With tractorisation getting in vogue, 9/11 tyne tillers with reversible shovels in alluvium and alfisols and duckfoot type in vertisols, tractor-drawn mouldboard and disc ploughs became popular. For secondary tillage trailing and mounted type tandem and offset disc harrows are in use and manufactured in large numbers through SSIs. However, versatility of 9/11 type tillers have made them most popular tillage implement with the tractor owners like that of Desi Hal in the past. In some places tandem disc harrows are widely used for seedbed preparation with occasional or no ploughing.

Research studies under ICAR All India Coordinated Research Projects on Energy Requirement and Farm Implements and Machines have established roto tilling as an efficient seedbed preparation equipment both under upland

and wet land situations. As a result it is currently under promotion with incentives in several states specially where vegetable crops are taken after rice. For wetland seedbed preparation power tillers with rototillers are very useful specially for small and marginal farms. However, many farmers who do not maintain bullocks, find custom serviced seedbed preparations fast and affordable. Many tractor owning farmers do custom servicing in seedbed preparation and seeding and in transport operations to improve economics of tractor owning.

Irrigation and Drainage Equipment: Mechanisation in lifting of the irrigation water was the first important land mark in modernisation of Indian agriculture when in late 1950s and early 1960s the efforts to adopt engine and electric motor operated centrifugal pumps led to indigenous manufacture of small diesel engines (5 hp) and 3-5 hp agricultural motors, and matching centrifugal pumps by mid 1960s. It expanded rapidly and today about 0.90 million irrigation pumps are manufactured annually, one-third engine operated and two-third electric motor operated. With over exploitation of GW and water table going down, the use of submersible pumps is on rise.

Solar- photovolteric water pumping sets are also available which at this point of time involve high capital cost but operating cost is very low. With irrigation waters getting scarce and costly their efficient application has become essential. As a result the pressurised irrigation systems—sprinklers, microsprinklers, and drip systems have come in use that save about 35-50% of irrigation water with 25-35% increase in the yield. Through fertigation drips improve the fertiliser use efficiency also. Efforts are on to improve the water use efficiency in Indian agriculture. In ricewheat cropping system, zero-till drill and raised bed planting are found to reduce water requirement about 35-50% with about 5-10% yield advantage.

Surface and subsurface drainage systems, and vertical drainages have been designed, installed and studied in actual farming conditions which have facilitated crop cultivation in otherwise water logged areas and increased yields substantially where water tables were close to the root zone or soils are saline or acidic due to adverse geo-hydrological situations preventing yield potentials being realised. Low-lift high discharge pumps have been developed for vertical drainage. Command areas of several medium and large irrigation or multipurpose projects have been studied and remedial areas evolved to improve water distribution and productivity through better water management. Energy audits, however, in many of the irrigation and drainage installations can not be considered favourable. There is considerable room of improvement of system efficiency through improved design, manufacturing, installation and upkeep of these systems tackling corrosion and encrustation problems.

Sowing and Planting, Transplanting and Checkrowplanting: Sowing crops in lines so distanced apart that farm power units can move in inter-row spaces without damaging the plants is one of the perquisites of mechanisation of post-seeding operations. As a result drills and planters have become popular. Placement of band of fertiliser by the side of seed slightly deeper is found to have better plant response and more yields. To realise this advantage, seed-cum-fertiliser drills succeeded the seed drills. For best results it is not only desirable to have seeds in row but also certain distance between the seeds (plants); to achieve it, there are planters plate type, cupfeed, and pneumatic seed metered but not as popular as drills.

There is talk about Madagaskar rice cultivation technology whose main features are checkrow transplanting of young rice seedling (15-20 days old) and interculture in standing crop along and across the rows. However, we do not have yet a mechanical check-row transplanter. Transplanting of seedlings/saplings is also widely practiced in fruit and vegetable and cultivation, agro-forestry where mechanisation is needed for which effort are on for some of the operations mechanised options have been created like pit making.

Interculture and Weeding Equipment: Removal of weeds, aeration of the root zone, creating soil mulch, and earthing are major objectives in interculture. In some cases thinning and gap filling or band placement of fertiliser are also done during interculture. Traditional interculture tools are Khurpi or Kudali which are very slow and unaffordable at current wages. As an improvement wheel hoes have been developed, some with several attachments to perform various interculture jobs. These are becoming popular. There are tractor mounted cultivators, however, due to absence of headlands, failure to create straight rows and other reasons these are not widely used. For wide-row sown crops like maize, pigeonpea, groundnuts, power weeders have been developed and commercialised. Self propelled rice power weeders have been developed but not popular yet. There are urea supergranule and prilled urea applicators developed for rice that improve fertiliser use efficiency.

Plant Protection Equipment: For applying weedicides, insecticides, fungicides as well as certain growth regulators spraying and dusting is followed. India has manual, knapsack auxiliary engine operated, manually pushed auxiliary engine operated, power tiller & tractor mounted and stationary engine operated sprayers and dusters in use, indigenously manufactured mostly in private sector. Annual production of sprayers and dusters is about 0.4 million. Recently Aeroblast sprayers, an introduction from Brazil has found good acceptance for cotton and tall tree applications. Manual and power operated sprayers could be single lance or long boom swath type, the latter are better for field crops and the former for spot treatment, kitchen gardens and trees.

Impingement, distribution, coverage and drift are major issues in spraying and dusting which can be addressed by electrostatic spraying and dusting. However, India lack oil based pesticides required for electrostatic spraying. Air assisted sprayers and dusters improve the coverage which is essential when using contact type pesticide. There are already a few in the market. Selection of right kind of nozzles is vital in achieving efficacy. R&D efforts continue to be made in this direction.

Harvesting, Threshing and Combining: Timely harvesting, separation of the produce and residue i.e. threshing and cleaning are very basic unit operations to save the crop from the qualitative and quantitative losses. Delayed harvesting means over maturity, and loss of quality, shedding in field, weather damages, damage due to fire and theft etc. Sickle, sythe, axe, Kudali, Gandasa, Dau are traditional harvesting equipment. Improved serrated sickles are available in the market at affordable price which save time and efforts. Sickle and sythe action are now mechanised with the introduction of walking, riding and tractor mounted vertical conveyer reapers.

Power threshing, replacing the tiring and slow bullock treading or threshing with flail or sheaves striked on hard surface, got introduced in 1960's. Wheat power thresher developed at Allahabad Agricultural Institute, Allahabad revolutionised not only the wheat threshing but set a process in motion that today we have power threshers specialised and multi-crop handling all cereals, pulses and oilseed crops. Traditional threshing of wheat and several other crops is totally replaced by power threshing where current trend is for high capacity bulk-fed power threshers. Though the power threshers have feeding chutes designed to avoid hazards to operators but some people remove them and under fatigue or toxicants get injured. It brought a legislation on farm machinery safety. Nevertheless injury due to farm equipment continue to exist. Reaping with VCR and power threshing optimizes harvesting, threshing, residue management, and feed recovery system.

Combining is found to be the cheapest method of harvesting and threshing rice and wheat, as a result combining has become popular and there are about 30 manufacturers manufacturing combines—tractormounted, tractor driven, and self-propelled with 8′-14′ headers. Standard combines, however, do not handle crop residue which have feed value and create residue management problem. Straw combines do provide a solution however, in about 60-70% combined fields straw is insitu incinerated to clear the field for the next crop. Many crops specially fruits and vegetables, tuber crops continue to be cultivated and harvested in the traditional way but for groundnut and potato diggers that have been developed and commercialised.

There are improved fruit pluckers, coconut tree climbers, manual and power operated coconut and arecanut dehuskers developed and commercialised. With increasing labour wages, conventional harvesting and threshing have become unaffordable as a result rising demand for mechanised harvesters specially of F&V and commercial crops for which solutions may not be easy to come. Efforts have been started to adopt appropriate harvesting machinery in use in developed world, however, most of these are high capital cost machinery which can be used on custom hire basis or as part of contract farming. There is demand for cotton pickers and sugarcane harvesters for harvesting commercial crops—cotton and sugarcane.

Transport Machinery: Traditional transport of carrying on head, shoulder or shouldersling or even as animal pack have lost relevance except where material to be transported are in very small quantities and for a short distance. Traditional bullock, camel or and horse carts are good for about 1-1.5 t of pay load. Pneumatic wheeled bullock carts can pull up to 2-3t of payload at 2.5-3.5 kmph and are still found economical for 5-10 km lead distance. Tractorised farms are seen keeping a pneumatic wheeled bullock cart for on-farm fodder, farm produce and input movements. Tractor owning farms have two wheeled 2-3t, trolley or 3-5t four wheeled trollies for on-farm and off-farm transport. In many cases 30-50% times agricultural tractors are used for transport of building construction materials improving annual use of tractors for economic sustainability.

There are hydraulically tilting trollies available for automatic unloading of bulk loaded materials. Majority of the trollies are back dropped and some back and side dropped type too. Carts and trollies are multipurpose transport vehicles used for movement of men and materials like seed, fertiliser, soil, manure, agricultural produce and residues. There is need for animal and power operated manure transporter-cum-spreaders. However these are not available yet in the market. There are specialised carts and trollies developed for movement of men in rural areas but not popular yet.

Dehusking, Shelling and Decortication: Dehusking, shelling and decortication are important for on -farm value adding and primary processing to make the produce more presentable to the buyer, and help in reducing transport, handling and storage costs before further processing and utilisation. Maize cobs need to be dehusked and shelled to get grain for marketing, processing and use.

Coconut, arecanut are dehusked dried before bringing in the market. Groundnut, castor, almonds etc need to be decorticated to get the kennels. Low cost manual and power operated dehuskers, shellers and decorticators have been developed. For minor oilseeds like cucumber seeds, neem, mango, kusum etc. decorticators have been developed or adapted but yet to be fully commercialised.

Cleaning and Grading: Separation under natural air stream, cleaning with cleaning basket ('Supa') and sieving with manual sieve and manual grading have been in vogue to remove chaff and foreign matter and quality improvement through grading using sieve in case of foodgrains and oilseeds. In case of F&V manual washing and grading are generally practiced.

Improved winnowing fans, winnowers, pre-cleaners, cleaners and graders both manual and power operated have been developed and adopted, commercialised and used for foodgrain and oilseeds, seed. Power operated washers, graders and packaging lines have been developed for F&V

and many of them commercialised but not popular yet. Seed cleaners and graders achieving very high degree of efficiencies (over 98%) are commercially manufactured and in use.

Drying and Dryers: Drying under sun or shade where hot dry air conditions prevail is a common traditional method of drying of foodgrains, fruit and vegetables and many other commodities to extend their shelf life for their safe storage for off season use and have value added products for additional income. However under unfavourable condition of hot, humid and overcast or rainy weather drying gets too slow so much so that before drying is accomplished molds and fungi develop contaminating and spoiling the material.

To overcome this handicap different kind of drying methods and dryers have been developed but not in use except for the milling units or home level solar dryers. In North India, rice and wheat quality is being adversely affected due to untimely rains and overcast weather, and inability of the farmers to handle safely the moist stalks. There is a case for mechanised combining and drying to save the foodgrains and oilseeds from weather damages, discolouration, attack of microorganisms etc.

Storage: Consumers of about 65% of the agricultural commodities produced are rural people who retain cereals, pulses, oilseeds etc for family consumption and dispose only surpluses (on average about 35%).

Traditional storage practices and storage structures work satisfactorily under favourable condition. However, under scientific storages/ warehousing losses to foodgrains are below 3-4%, but under on-farm storages losses are about 10% or even more as in the case of legumes. As a result farmers tend to dispose off vulnerable surplus even under depressed market situation thereby suffering economic losses.

Improved storage structures and practices have been developed and some of them have become popular also like metallic bins, use of tunnage to store foodgrains in bags or bulk for short term, drying to safe moisture levels and other prophylactic measures before storage, use of fumigation (phosphene tablet) in case of insect infestation etc.

For storage of certain perishables use of evaporative cooled storage/zero-energy cool chambers are found useful. Public and private sector cold storages, controlled atmosphere storages, onion storages are being used for perishables and the Govt of India has launched a scheme to promote cold stores and onion storages. Scientific on-form storage can bring down losses dramatically at the same time enable the farmers to sell their produce off season at better prices.

On-Farm Milling and Processing: Indian villages have traditionally been producer-cum-primary processing centres. To enable the rural sector to meet its needs at least cost and market value added products for additional income and employment, processes and equipment/pilot plants have been developed like mini-rice mills, mini-dal mills, mini-grain mills, mini-soy milk-cum-tofu plant which are commercially available.

High performance mini-oil expellers have been developed and commercialised enabling growers crushing and filtering oilseed for fresh wholesome vegetable oil for own consumption, marketing in the nearby towns and retaining cake for the livestock or sale to solvent extraction units. Models of Agro-Processing Centres have been developed.

Mechanisation Policy and Legislative

Mechanisation Policy: Agricultural mechanisation is a relatively less regulated sector. Increasingly need for agricultural mechanisation have been demanded and discussed in legislative and policy making bodies. Perceptions vary but the most common description of agricultural mechanisation needed in India is appropriate selective mechanisation which brings economic competitiveness and removes avoidable drudgery.

National Commission on Agriculture (NCA) made stipulations on agricultural mechanisation. High powered committees appointed by GOI have given direction the subject. National Agricultural Policy (NAP) amplifies agricultural policies. A draft agricultural mechanisation policy has been worked out jointly by Machinery Division, DAC, GOI, which needs to be nationally agreed and adopted.

Standardisation & Certification: To protect the farmers' interests from introduction of inappropriate trouble some equipment in Indian market, GOI has requirement of testing and certification by DAC which is carried out its Testing and Training. As a mark of standardisation and quality BIS standards are there on farm equipment, ISI certification on critical components and machines.

There is a long list of such standards. BIS standards are steadily getting closer to ISO where available. There are farm implements and attachments to tractors reserved for SSI which helped in bringing down the unit cost of these equipment. However, the quality suffered. To get over this crisis following equipment have been dereserved.

— Crow bars
— Levellers
— Mowers
— Reapers up to 5 hp
— Seed Drills up to 5 hp
— Other agric machinery up to 5 hp

Gender Bias in Mechanisation: The National Commission on Women is exercised putting pressure on R&D organisations and development departments to remove drudgery to farm women and to have hand tools, implements and machines suited to different sections of the farm women. These are receiving attention. The following items have been identified for due attention on priority.

- — Improved sickle
- — Wheel hand hoe
- — Dibbler
- — Maize sheller
- — Groundnut decorticator
- — Horticultural tools
- — Sugarcane stripping knife
- — Cleaners
- — Graders
- — Improved chulhas
- — Solar cookers
- — Husk fired stoves
- — Manually operated seed-drill/planter
- — Manual fruit harvester
- — Cotton stalk puller
- — Dal mill
- — Chilli seed extractor

Demand due to Globalisation: Farmers' interest groups are putting pressure through different fora to mechanise harvesting of commercial crops like sugarcane, and cotton. Demand is also raised to have mechanised tea harvesting.

Globalisation has also put special demand on Indian equipment industry to improve quality, durability, replaceability of fast wearing parts and economic advantages—capital cost as well as unit cost of operation.

Dangerous Machine Act 1983: The Industries (Development and Regulation) Act 1951, Factories Act and Workmen's Compensation Act 1923 were examined for the purpose of prevention of accidents on the threshers. Based on the above Acts, the Dangerous Machine Act was passed and promulgated by GOI in 1983. This Act was passed following a series of accidents involving the loss of limbs in agricultural operations.

The Act provides for the regulation of trade, commerce, production, supply, distribution, and use of the products of any industry producing dangerous machines, with a view to securing the welfare of labour operating such machines through a system of licensing of the manufacture of such machines operated by a Controller appointed under the Act.

The employer is to take out an insurance cover to pay compensation to the operator of the dangerous machine in the event of injury or death. The salient features of this Act are:

- Manufacturers should ensure that every part of dangerous machines confirms to BIS standards.
- Duty of manufacturers to supply operators manuals with each dangerous machines.
- User to get dangerous machine registered.
- User to get the existing dangerous machine modified.
- Employers liability for compensation.
- Notice of accident.
- Open insurance policy of the operator.
- The bill authorities inspection and seizure of the machine causing death or injury.

As a result of the Dangerous Machine Act the scientists designated safe feeding chutes for different type of threshers keeping in mind the anthropometric dimensions of the upper limbs so that they could not be pulled in to the thresher while feeding the crop in the thresher.

Indian standards related to safety

The following standards on safety have been formulated by the Bureau of Indian standards:

- IS:12239(PT1):1996 Guide for safety and comfort of operator of agricultural tractors and power tillers: part 1 General requirements (first revision)
- IS:12239(PT2): 1988 Guide for safety and comfort of operator of agricultural tractors and power tillers: Part 2 Requirements relating to agricultural tractors

— IS:12239(PT3):1988 Guide for safety and comfort of operator of agricultural tractors and power tillers: Part 3 Requirements relating to Power tillers
— IS:8265:1996:Agricultural tractors - Guards for power take -off (PTO) drive shafts (second revision)
— IS:4931-1995:Agricultural tractors - Rear mounted power take off types 1,2 and 3 (third revision)
— IS:5994-1987 Test code of agricultural tractors (second revision)
— IS:6024:1983 Guards for harvesting machines (first version)
— IS:9019:1971 Code of practice for installation, operation and maintenance of thresher
— BIS:6283:1971Symbols for operator controls on agricultural tractors and farm machinery
— IS:9020-1971 Safety requirement for power threshers
— IS:9129:1979 Technical requirements for safe feeding system for power threshers
— IS:9581:1988 Safety and operational requirements for pedestrian- controlled cylinder(reel) power lawn mower
— IS:9632:1988 Code of practice for operation and preventive maintenance of crop protection equipment
— IS:1061:1983 Pictorial representation for cautionary notices for power threshers

Agricultural Mechanisation Trends

Agricultural mechanisation trends are linked with the trends in agriculture, agro-processing and rural living, globalisation of world markets and market trends, WTO obligations, and State and Central Government policies and demands of political constituents.

Agriculture is a state subject. Decisions at state or regional level selfsufficiency or policy to concentrate on exploiting agroecological advantages and meeting short falls through imports from other states of the Union or foreign source are likely to affect the mechanisation trends. As of now things are in the state of flux.

Country is faced with contradictory situations each having its own requirements including mechanisation. Moder-nisation requires sophistication in mechanisation which is possible at relatively large scales of operations entailing capital and management constraints. Globalisation puts heavy demand on competitiveness, reduced unit cost of production, indirectly demanding mechanisation and to a certain extent automation.

These will lead to tractorisation. Marginal and small farmers are increasingly becoming part time, absentee farmers, periurban farmers, wage earners on part or full time basis. Industry and service sectors, trade and commerce unable to reduce land based livelihood compel the rural people to remain on land based livelihood, forcing steady increase in number of land holdings but average land holdings going down making mechanisation more challenging and difficult.

Scaling down of farm machines reduces mechanical advantages. Instead of owning farm machinery other than hand tools such marginal farms can meet their needs through custom servicing, if it is well developed.

— Country is faced with the basic livelihood issue of the rural masses. With per capita arable land availability dwindling and average land holding coming down to levels that it is too difficult for the farm families to have minimum acceptable standards of food, shelter, clothing, health care, and education, the Central and State Governments are seized with the issue of ways to widen livelihood base of these people. Increasing productivity; crop diversification towards horticulture, livestock husbandry, fishery and forestry; post-harvest management for minimisation of post harvest losses, value addition and agro-processing for additional income and employment are some of the developmental measures enunciated and related schemes and programmes launched.

— DAC, Government of India have launched Agriclinic and Agri-business Scheme that is going to supplement and substitute some of the extension functions and services rendered by the public sector extension. It is envisaged that entrepreneurs will be providing custom service to the farmers besides other services which will need machines of large capacity. The entrepreneurs will be also taking agro-processing, packaging, transport and marketing activities both for domestic and foreign markets.

— Governments are also encouraging contract farming i.e. farmers producing specified agricultural produce needed as raw material for industries. Governments are also reducing intermediaries between growers and consumers. Growing middle classes where both husband and wife are working need processed and semiprocessed, ready to cook, ready to eat items.

— Under such complex scenario the emerging trends in mechanisation of agriculture, agro-processing and rural living are as follows:

Role of Farm Power Units

— Animate energy, both human and draft animals (DAP) will continue to be important sources of farm power specially on small and marginal farms, hill agriculture, horticultural crop cultivation, animal husbandry and fishery. However, use of DAP is likely to decline slowly and may stabilise around 20-25 million pair of bullock equivalent. Due to lack of opportunities in other sectors of Indian economy, use of human as farm workers is likely to grow but at a slow rate with elements of under employment.

— There is going to be increasing demand for riding type farm equipment. Animal-drawn tool carriers under different brand names are going to get in use. Likewise riding type power tiller farm operations will be in demand. Some power, no doubt is lost in transport of the operator and the tool carrier, but the loss is more than made up through gain in net operational time due to favourable work-rest cycle in a riding type manmachine system. Drudgery to the operator is minimised.

— Small engines (1-3 hp) for hand-held and knapsack power operated equipment for pruning/coppicing, hedge trimming, cutting of bushes, tree felling, tea harvesting, spraying, and dusting etc are likely to become popular.

— Due to global competition, agricultural engines-diesel, petrol, kerosene, biofuel are going to be better through superior metallurgy and manufacturing processes. Advances in combustion chamber designs, fuel injection etc shall make them more fuel efficient. Noise and vibration problems are going to receive greater attention.

— Paucity of petroleum reserves is going to bring biofuels - alcohol from sugarcane, molasses, corn cob etc and non-edible vegetable oils certified for use as biodiesel.

— For stationary operations for shaft power, electrical power generation, and process heat for agro-processing alternate energy sources such as producer gas from crop residues, farm and roadside grown energy plantations, processing wastes like rice husk, grou-ndnut shell etc are going to find greater acceptance. Solar Thermal System and Thermal Power Units run on crop and processing residues are likely to gain popularity as stand alone power units.

— Light weight and modular PT for hill agriculture/ terraced farming are going to appear using diesel and biofuels.

Farm Implements and Machines

- Improved energy efficient matching implements and machines for different unit operation of agriculture are expected to be available soon.
- For timeliness, efficacy and reduced unit cost of operations single-run machines are going to gain acceptance.
- High ground speed machinery for tillage, sowing, and planting using rolling, rotary and vibratory actions.
- Custom hiring in seedbed preparation, sowing, planting, transplanting, harvesting and threshing and other specialised unit operations where ownership of a costly equipment is not justified.
- Microprocessor controlled equipment for land grading, seeding, transplanting, and farm machinery manag-ement.
- Trade in used farm machinery, repair, reconditioning is likely to come up in order to reduce capital requir-ements of farming.
- General improvement in quality of farm machinery compelled by global competition, export prospects through R&D, standardisation, and quality control measures and advanced manufacturing methods.
- Use of plastics in farm machinery manufacture specially for hoppers, handle grips, conduits and tubings, metering devices, surfacing coating and components requiring resilience or corrosion resistance etc. Advances in polymer chemistry and precision in injection mouldings can turn out cheaper nozzles and control valves, gear drives.
- Components and whole equipment designed with ergonomic rigour conforming to anthropometry of workers, and operated on scientifically established work-rest cycles are going to be there. Working postures are likely to change from bent and sitting to

erect postures for greater work output and less strain on body

— Conventional tillage and sowing are going to be supplemented and substituted by rotary tillage, conservation tillage, raised-bed systems for economy, efficacy, and enhanced productivity.

— Conventional Biasi, Lehi and manually transplanted rice are going to be supplemented and substituted by mechanical transplanted rice using manually operated and self propelled transplanters and mat type nursery. Dry drilling and paddy sown with pregerminated paddy seeder in levelled puddled field (Lehi substitute) have already made in-roads. Check-row transplanted and bidirectional mechanically weeded paddy crop has yield advantages and thus likely to gain ground provided affordable check-row planters and efficient weeders are developed and made available.

— Zero-till drill and raised bed planters in rice-wheat ecology are going to become still more popular for the economy in use of irrigation water (35-50 saving), reduction in unit cost of production (1500-2000 Rs/ha), better control on phalaris minor and yield advantages (about 5%).

— Manual, animal drawn, power tiller and tractor mounted and self propelled power weeders for wide row crop as well as rice-wheat are going to be in greater use and would be available in wide range.

— Sugarcane planters; cutter-planters which makes sets, plants, applies fertilizer, weedicide and pesticide in one go are going to gain popularity.

— Sugarcane stubble shaver and ratoon management equipment are going to be in demand.

— Drills and planters for direct seeded vegetables, vegetable transplanters for potted and soil block mature seedlings of brinjal, chillies, cole crops, onion, garlic etc are expected to be in use as vegetable cultivation gets commercialised.

— Polyhouse nursery-conventional, potted, and soil block as well as semi-automatic nursery are going to come up.

— Pressurised irrigation systems-sprinklers, microsprinklers drips, fertigation, skiprow, irrigation for water conservation and enhancement of water use efficiency, and associated yield advantages. Onfarm agricultural drainage-surface, sub-surface and vertical are going to be integral part of onfarm water management. Drainage not only improves productivity but also influences such factors as lodging of the crop along with seed rate and N-application. Mechanisation in laying drainage system is going to gain grounds.

— Wide boom and high ground clearance sprayers, tall tree sprayers and dusters, turbofan power sprayers and dusters/aeroblast sprayers and dusters will come in greater use. Environment pollution associated with spraying and dusting and pressure being mounted by environmentalists is likely to give rise to use of electrostic spraying and dusting which has better impingement and very little drift.

— Sickle harvesting of cereals, specially rice, wheat etc is being substituted by reaping with vertical conveynor reapers - walking, riding, and PT & tractor mounted. These equipment will be in greater demand.

— Power threshing is already in vogue for most of the crops through owned or custom hired power threshers. Trend is emerging in favour of large capacity, bulk fed, multicrop threshers. There are going to be power threshers difficult to thresh for crops like pigeonpea.

— Use of standard grain combines for large number of cereals, pulses, and oilseeds are already there growing

@ over 10% annually. Demands for grain combines, plot combines is going to rise.

— Harvesters for commercial crops like sugarcane, cotton are going to be there in not too distant future. Forage harvesters and forage combines and hay making equipment are going to be in demand.

— Agri-electronics has emerged as an area fro R&D and going to make available a number rugged, user friendly, affordable electronic gadgets to practice precision farming like pH meter, water quality analyser, macro and micro-nutrient analysers, machine vision for plant growth management, analysers for determination of quality of produce electronic ultrasonic pest repellers,. electronic bird scarer etc are some of the possibilities. Robotics in fruit and vegetable harvesting have already been field tested in developed world where labour is very expensive. Automatic irrigation systems are in use already in India at pilot level.

— Organic farming gaining momentum is going to have improved equipment for encorporation of biomass in the soil, FYM and compost spreaders. Slurry pump, tanker and slurry applicators are also going to be in demand to recycle solid and liquid wastes of livestock. Mechanised composting has now better chances of acceptance.

Agro-Processing

— On-farm produce management is going to receive greater attention to minimise harvest and post harvest losses from qualitative and quantitative damages. This entails harvesting at right time, using right harvesting tools, implements and machines, handling of the produce in a manner that mechanical, weather and physiological damages are minimised specially in fruits and vegetables.

— On-farm value addition for additional income and employment and ease of marketing and retaining processing and milling by-products of feed and fuel value on the farm itself at the same time saving in handling, transport, storage, packaging costs etc. This is going to put dehuskers, shellers and decorticators, cleaners and graders in greater use. There are going mechanical washes and fruit and vegetable packaging lines.

— Agro-processing to meet local needs of rice, atta, dal, oil, ground spices, fruit and vegetables and livestock feed, concentrates using mini-grain mills, mini-modern rice mills, mini-expellers and batch solvent extraction units, hammer mills, mixers, bagging and stiching units etc are going to be there in production catchments.

— Urban middle classes are looking for ready to cook even ready to eat items. This is likely to give place to soy-milk and tofu plants both in rural and urban areas. Minimally processed vegetables and fruit juice concentrate, jam jellies, candies etc can be produced in production catchments and supplied directly to retailers at much lower consumer prices. Such processing units are going to come up.

— Scientific storage avoiding distress sale and providing off-season supply is a value addition activity. Warehouses, silo storages, storage in evaporative cool chambers, cold storage, onion storages have good prospects and likely to come up in a big way.

— Drying is an age old method of preservation of agricultural produce and to have long shelf life food products from perishables, mostly sundried. Heated air dryers-batch type, batch-in-bin, recirculatory, and continuous type are there for foodgrains and such particulate materials, some of these are in use at millers level already. Onfarm mechanical drying is likely to come up to save moist grains from discolouration and from microbial damages. Tray dryers are widely used for drying fruit and vegetable slices. For on-farm drying of fruits and vegetable solar-tray dryers have been developed and are gaining acceptance. Osmodely-drated fruit slices are very appealing and organoleptically rated high. Scraped surface dehydrators are likely to come to concentrate milk in to Khoa and other such applications.

— Agro-Processing Centres appropriate for different production catchments using modern processing equipment matching to the scales of operation are going to come up enabling producers meeting their own needs at the least cost at the same time market surpluses after value addition.

— To handle, storage, transport, and market perishables cold-chains are needed. Available ones are too expensive. Indigenous cold-chains are going to be there for different commodities. In a way it exists for milk, fish, meat etc. However, greater innovation and creativity is expected in this area.

— Processes and pilot plants are expected to be producing speciality diet foods like soy-milk, tofu, soy-fortified biscuits, soy-yoghurt, soy-icecreams, tempeh bergers etc.

— Processes and pilot plants are going to come for powder and liquid jaggery, beverages from sugarcane and traditional fruits.

— Particle board and timber substitutes made out of jute and kenaf sticks, paper and pulp from jute cuttings, sticks, and whole plant; craft paper and corrugated board packaging boxes out of cotton stalks are demonstrated already more commercial units are likely to emerge.

— Mini-jute spinning system developed at NIRJAFT on the pattern of khadi making utility product from jute fibre and their blends holds promise.

— Cotton ginning and baling is being modernised under Technology Mission on Cotton which is going to put in place high efficiency roller gins, sawgins, precleaners, electrically controlled balers, and mechanical handling systems in modernised ginneries which suit cotton trade and textile industry.

- Processes and pilot plants are going to come producing seed lac, shellac, bleached lac, lac wax, paints and varnishes, edible grade lac dye, valuable chemicals from lac and its by-products.
- Jute and textile industry are going to experience greater modernisation and automation.
- Jute and kenaf ribboners and decorticators to enable ribbon retting for quality fibre and minimise pollution are going to come in use.
- For profitability and sustainability of commercial crop thrust on diversified uses is likely to continue demanding development of new processes and equipment to realise the prospects.

Rural Living

- In a village eco-system, 80% of the total energy spent goes to domestic sector and of that 80% goes for cooking. Fetching of drinking water, collection of fuel wood are ordeal to rural women and children. Rural kitchens are full of smoke, illilluminated, cause of many health hazards. Quality of fuel decides quality of life to a housewife, rural or urban alike. In that scenario biogas stoves, pyrolysed briquetted fuel and sigdi, biogas lantern, biogas run ovens; solar cookers—individual and community type, solar water heaters are under promotion and are likely to increase in number supplementing and substituting conventional cooking and water heating.
- With the rural electrification in India, refrigerators are reaching rural homes which provide capacity to store perishables safely and prevent wastage.
- Pressure cookers which are energy efficient are also

reaching rural homes.

— Other kitchen appliances like mixer grinders, electric iron, electric fans and lights are being used in electrified villages.

— Butane supply to rural areas is gradually spreading which is going to conserve fuel wood and other biomass for organic recycling.

— Smokeless cooking stoves that are energy efficient therefore, under promotion in a big way.

— Hapur Kothis, metallic air tight storage bins have found acceptance that keep the food grains safe protected from rodents, and enable fumigation in the event of infestation.

— Ata chakkis are doing custom milling likewise motorised wet grinders have become popular reducing drudgery in rural home. Single phase electric motor run cleaning grading and size reduction equipment also open avenues of custom servicing to other.

— Rural water supply and sanitation and rural roads and public transport are gradually growing. Dry and wet type bore hole latrines provide cheap sanitation to rural homes

— Biogas plants, composting, vermicomposting allow rural people to convert organic wastes in to wealth extracting energy, and plant nutrients in litter free, fly free, incineration free manner.

— Rural agro-processing centres and other home level agro-processing and craft activities using mini-burr mill, solar dryers etc are enabling additional income and employment in spare time.

References

Alam, A. and Ilyas, S.M., *Agro-Processing for Additional Income and Employment*, National Symposium on Empowerment of Women in Agric, RAU, Pusa (Bihar), 4-6 March, pp 1-16, 2001.

_______________ , *Post-Harvest Management of Agricultural Produce and By-products*, Paper, Agric Sci Congress on Sustainable Dev of Mountain Agric, AAU, Guwahati, 4-7 April, pp 1-28, 2001.

Anon, *Zero Tillage Technology - Wheat production at Lower Cost*, DWR, Karnal,. 2001.

_______________ , *Electronic Ultrasonic Pest Repeller*, Dhakar Engineers, Pratapnagar, Udaipur.

Bisht, B.S., Kachru R.P. and Alam, A., *Potential for Small Scale Agro-Processing Industries in Rural Areas of Chhattisgarh*, Paper presented in launching workshop on Raipur Centre of AICRP-PHTS, 14aD, Raipur, 10 Sept, pp 1-21, 2001.

Cowx, I.G., *Developments in Electrical Fishing*, Oxford Fishing News Books.

Gite, L.P. and Singh, G., *Ergonomics in Agricultural and Allied Activities in India*, Tech Bulletin CIAE/97/70, CIAE, Bhopal.

Khan, A.U., *Dual-mode All-Crop Thresher for Egyptian Conditions*, AMA 21(4): 11-14, 1990.

Pandey, M.M., Majumdar K.L., Singh, G. and Singh, Gajendra, *Farm Machinery Research Digest*, Tech Bul CIAE/ 97/69, CIAE, Bhopal, 1997.

Ozmerzi, A. and Barut, Z.B., *Mechanization Level in Vegetable Production in Antahya Region and Turkey*, AMA 29(1): 43-46, 1998.

Bibliography

Alam, A. and Ilyas, S.M., *Agro-Processing for Additional Income and Employment*, National Symposium on Empowerment of Women in Agric, RAU, Pusa (Bihar), 4-6 March, pp 1-16, 2001.

Asian Productivity Organization, Farm Mechanization in Asia, APO, Tokyo, 1983.

Baig, M.A., The Tractor in India - A significant instrument for future development, Proc. of the Symposiumon Farm Mechanization - Problems and Prospects ISAE, North Chapter and ISAE, PAU, Ludhiana pp. 57-66, 1978.

Balasankari, P.K. and Salokhe, V.M., *A case study of tractor utilization by farmers, Coimbatore district*, India. Agricultural Mechanization in Asia, Africa and Latin America 30(3), 14-18, 1999.

Balishter, Gupta, V.K. and Singh, R., Impact of Mechanization on Employment and Farm Productivity, Productivity, 32 (3): 484-489, 1991.

Bisht, B.S., Kachru R.P. and Alam, A., *Potential for Small Scale Agro-Processing Industries in Rural Areas of Chhattisgarh*, Paper presented in launching workshop on Raipur Centre of AICRP-PHTS, 14aD, Raipur, 10 Sept, pp 1-21, 2001.

Brar, S.S., *Dynamics of agricultural production pattern under varying input output price in Punjab*, Unpublished Ph.D. (Agril. Economics) thesis, PAU, Ludhiana, 1998.

Chopra, K., Tractorization and Its Impact on the Agrarian Economy of Punjab, A Report submitted to USAID, American Embassy, New Delhi, 1972.

Cowx, I.G., *Developments in Electrical Fishing*, Oxford Fishing News Books.

Desai, D.K & Gopinath, C., Impact of Farm Tractorization on Productivity and Employment (Gujarat State) Centre for Management in Agriculture, IIM, Ahmedabad, 1975.

Dhillon, B.S. and Sidhu, D.S. "Economics of custom hiring of agricultural machinery in Punjab – a few case studies", Agricultural Engineering Today, 11(2) 15-19, 1987.

Gite, L.P. and Singh, G., *Ergonomics in Agricultural and Allied Activities in India*, Tech Bulletin CIAE/97/70, CIAE, Bhopal.

Grewal, S.S., *Impact of mechanization (Tractorization) on labour employment and labour productivity in Punjab*, Unpublished Ph.D. (Agril. Economics) thesis, PAU, Ludhiana, 1972.

Hossain, M.A., *Estimation of demand for major agricultural inputs in Punjab agriculture*, Unpublished Ph.D. (Agricultural Economics) thesis, PAU, Ludhiana, 1988.

Joshi, H.C., *Pattern of tractor power utilization in a fodder farm: A case study*, Agricultural Mechanization in Asia, Africa and Latin America, 29(3). 39-41, 1998.

Khan, A.U., *Dual-mode All-Crop Thresher for Egyptian Conditions*, AMA 21(4): 11-14, 1990.

Kumar, N. et al., *Farm mechanization and its impact on Punjab agriculture*, paper presented in I.S.A.E. Seminar report, Khalsa College, Amritsar, 1995.

Lekhi, R.K. and Singh, Joginder, *Agricultural Economics*,Kalyani Publishers, New Delhi, July 2002.

Mishra, P. and Sundram, *Some aspects of the economics of harvest combines in Punjab*, Economic & Political Weekly, 10(39), 1975.

Ozmerzi, A. and Barut, Z.B., *Mechanization Level in Vegetable Production in Antahya Region and Turkey*, AMA 29(1): 43-46, 1998.

Pandey, M.M., Majumdar K.L., Singh, G. and Singh, Gajendra, *Farm Machinery Research Digest*, Tech Bul CIAE/ 97/69, CIAE, Bhopal, 1997.

Pathak, A.S., *Farm mechanization and productivity*, Productivity, 23(4): 411-417, 1985.

PAU, Ludhiana, Directorate of Extension Education, *Package of Practices for Rabi Crops* in Punjab, 2002.